The History of the Comstock Lode

JOHN W. MACKAY

THE HISTORY OF THE COMSTOCK LODE 1850 –1997

Grant H. Smith

with new material by Joseph V. Tingley

Nevada Bureau of Mines and Geology
in association with the
University of Nevada Press

Reno   Las Vegas

To John W. Mackay

His name was constantly on people's lips—almost invariably with words of praise. Everything about him was distinctive; his modesty, his reserve, his unfailing kindness to old friends, his innumerable benefactions, his uprightness, and the simplicity and decency of his life. Riches did not corrupt or steal away his good name, but rather served as his means to further develop American resources.

Nevada Bureau of Mines and Geology Special Publication No. 24

University of Nevada Press, Reno, Nevada 89557 USA

Library of Congress Cataloging-in-Publication Data

Smith, Grant H. (Grant Horace), 1865–1944.
The history of the Comstock lode, 1850–1997 / Grant H. Smith ; with new material by Joseph V. Tingley.
p. cm. — (Nevada Bureau of Mines and Geology special publication ; no. 24)
Rev. ed. of: The history of the Comstock lode, 1850–1920.   c1943.
Includes bibliographical references and index.
ISBN 1-888035-04-8 (alk. paper)
1. Comstock Lode (Nev.)—History. 2. Gold mines and mining—Nevada—Virginia City Region—History. 3. Silver mines and mining—Nevada—Virginia City Region—History. 4. Mineral industries—Nevada—Virginia City Region—History. 5. Virginia City (Nev.)—social life and customs. I. Tingley, Joseph V. II. Smith, Grant H. (Grant Horace), 1865–1944. History of the Comstock lode, 1850–1920. III. Title. IV. Series: Special publication (Nevada Bureau of Mines and Geology) ; 24.
TN413.N2S55 1998
338.4'76223422'0979356—DC21
98-19653   CIP

03 05 07 06 04
4 5 3

Contents

Contents

Contents

Contents

Illustrations

FOREWORD

The Comstock Lode is nationally famous for its huge output of gold and silver and its stirring history in the pioneer days of Nevada and the West Coast.

The Comstock, with this glorious history as one of the world's greatest mining camps and producers of the precious metals, will always attract mining capital searching for low-grade ores or bonanzas. The Nevada Bureau of Mines and the Mackay School of Mines, jointly, are striving to be a leader in publishing and collecting material dealing with that great Nevada mining district.

In mining camps, however old, well-explored and deserted, hope of a revival never dies in the hearts of the old timers; and the new rising generations of mining engineers stand ready to put their increased knowledge of geology, mining, and metallurgy, or of changed economic conditions, against the obstacles that closed the mines down.

Much in the way of fiction and semifiction and facts has been published, but there has been lacking a comprehensive, chronological mining history of the Lode, extending over a half century with a progressive record of the development work carried out, the failures encountered, the bonanzas discovered, and the production records of the mines. This is factual material that both mining investors and engineers first seek when investigating the possibilities of a district or a mine within a district.

The late John A. Fulton, as director of the Bureau, over ten years ago sought an author qualified to write a mining history of the Comstock Lode, who was qualified by both his life experiences and his interest in the Lode to write such a history.

Such a person was Grant H. Smith, a mining attorney of San Francisco. His youth was spent on the Comstock in the bonanza days as an inquisitive youngster and young miner; then he taught school and studied law and was admitted to practice by the Supreme Court of Nevada in January 1890. His later life has been in close contact with the Comstock as attorney for a number of the mines. Being a man trained in collecting and analyzing facts, along with the strong interest and perception in the historical events that stirred men's beings and brought out both the heroic and base in their characters, he brought also to this work the judgment of mature years and an unfagging search for source material.

Ten years of research brought forth a voluminous manuscript covering not only the mining history, but also that of political and social history, and particularly the life history of the Comstock's outstanding character, John W. Mackay.

The present director of the Bureau is of the belief that following the present war there will be, as with the last world war, a great revival of gold and silver mining· with higher prices per ounce than at present. Desiring to accumulate valuable information for that date, he prevailed upon Mr. Smith to take from his manuscript that mining material most suited for a Bureau bulletin, and with the aid and advice of the Director to rewrite it in the form of this bulletin.

Enough of the absorbing personal histories of the prominent men of the Lode has been retained along with many rare illustrations to entice the average citizen of Nevada to read and realize the importance of the Comstock and of the mining industry in the life of the State.

The University of Nevada and the Bureau have issued the following bulletins dealing directly with the Comstock:

Vol. 3, No. 4—The Ventilating System at the Comstock Mines, Nevada, by George J. Young, 1909 (out of print).

Vol. 26, No. 5—The Mines and Mills of Silver City, Nevada, by A. M. Smith, 1932.

Vol. 30, No. 9—Geology of the Silver City District and the Southern Portion of the Comstock Lode, Nevada, by Vincent P. Gianella, 1936.

During the last two years the Bureau, with the aid of research by the Nevada State Writers Project of the W.P.A., has compiled from the files of the old Comstock newspapers and written chronologically, the "Individual Histories of the Mines of the Comstock," fifty-six properties in all. This voluminous material is in typewritten form for consultation at the State Library in Carson City and in the Bureau's office at the Mackay School of Mines.

In addition, there will be available at a future date as a gift to the Bureau, a detailed account of the operations of each of the principal mines covering hundreds of pages, compiled by Grant H. Smith in the course of his research work.

Within the last decade the United States Geological Survey has made a thorough restudy of the Comstock Lode, Dr. V. P.

Gianella of the Bureau aiding in the work. Due to the all-out war effort of the Survey, this new material remains unpublished.

The Mackay School of Mines has in its library a large collection of books concerning the Comstock, in its museum on display an unrivaled collection of Comstock ores and historical relics, and in its files, a great accumulation of maps, company reports, and correspondence files.

JAY A. CARPENTER,
Director.

THE HISTORY OF THE COMSTOCK LODE
1850-1920

CHAPTER I

Gold Canyon Placers Discovered in 1850—The Grosh Brothers—
The Comstock Discovered at Gold Hill—The Ophir Dis-
covery—The Early Locators—Snow-Shoe Thompson—"Old
Virginny"—John W. Mackay—The Ophir in 1859—Early
Transactions.

The story of the Comstock begins on a pleasant May morning
in the year 1850, when a wagon-train from Salt Lake City halted
for a few days rest on the banks of the Carson River among the
willows and tall cottonwood trees where the town of Dayton later
grew up.

A small stream flowing from a range of high hills seven miles
to the west entered the river at that point, and young William
Prouse, having a little time on his hands, thought to wash some
of the sand in a milkpan to see if it carried gold. To his surprise
a few small "colors" remained in the bottom of the pan after the
sand and gravel had been washed out. The discovery was not
thought important and the party went on. But, when delayed in
Carson Valley by reports of deep snow on the Sierras, John Orr
and Nick Kelly returned and spent two weeks in prospecting the
gulch down which the little stream flowed. They found a little
gold all the way up to the mouth of American Fork where Orr
dug a small nugget out of a crevice with a butcher-knife.[1] The
diggings were poor compared with those they expected to find in
California so the train proceeded. Orr, meantime, had given the
gulch the high-flown title of Gold Cañon, although it was neither
rich in gold nor a cañon. Throughout most of its length it is a
shallow sage-covered gulch like a thousand others in Nevada.
The only place where the walls come close together is at Devil's

---

[1]Orr tells the story in a letter printed in the *Alta California* of May 17, 1880;
Thompson & West *History of Nevada*, pp. 29, 30 (1881), Lord's *Comstock Mining and Miners*, pp. 12, 13 (1883). It is reported that gold was found in the
sands of the little stream in 1849; if so, it was not followed up.

The Sierra Nevada is referred to in this book as "the Sierras," in accordance
with old-time custom.

Gate (another overblown title) where a narrow reef of rock lies across the cañon.

The story of the discovery, greatly exaggerated, passed along the road and other emigrants stopped to gather a little gold.[2]   It was soon learned that men could win an average of $5 a day in the cañon during the working season—some of the more fortunate earning more—and from 1852 to 1855 about 100 men worked there part of each year.   Many disappointed placer miners came from California every spring to earn better pay than they could at home, among them Allen and Hosea Grosh.   By 1855 the best ground had been washed,[3] earnings fell to $4 a day, and thereafter the number of miners steadily decreased.   Two years later men could earn but $2 a day with rockers when there was water, that is, about two thirds of the year unless the season was very dry. In 1858 a number of the miners left for the newly discovered placer mines near Mono Lake.

In nine years of operation the Gold Cañon placers produced less than $600,000, perhaps much less,[4] and were practically worked out.   The few remaining miners had to find new diggings or leave the half-indolent, carefree life to which they had become accustomed, of earning enough during the spring and summer to keep them during the winter.

"Old Virginny," who was more resourceful than any of the other miners, discovered some fair placer ground in 1857 along the little stream near the head of Six Mile Cañon, which he and a few others worked out in two summers.   This was just below the future Virginia City.   The next spring two other miners, Peter O'Riley and Patrick McLaughlin, came up from Johntown and began to work a little higher up, thinking to win enough gold

---

[2] The *California Courier* of July 8, 1850, reprinted the following from the *Sacramento Transcript:* "The accounts given us a short time ago by Mr. Shinnaberger in regard to the discovery of gold in Carson Valley (Gold Cañon) are fully confirmed. A large number from Salt Lake are now congregated in the new diggings, where they intend to remain, though their original design was to cross over to California."

[3] It was rounded gravel only in part; much of the pay dirt was disintegrated quartz.

[4] Lord estimated the production at $641,000, based on too-perfect information given him. *Comstock Mining and Miners*, pp. 24, 63.   Henry De Groot, who lived in the cañon in 1859 and 1860 and knew all of the remaining miners, estimated the total yield at "between three and four hundred thousand dollars." *U. S. Mineral Resources* for 1866, p. 87.

Nothing is known of the earnings of the large number of Chinese who worked for several years on the flats below the mouth of the cañon.

to outfit them for a trip to the Mono diggings. The ground was poor and they were almost ready to quit when they dug into a layer of rich black sand that proved to be a concentrate from the top of the hidden Ophir bonanza.

In January 1859 "Old Virginny" persuaded three of his friends at Johntown to go with him to the head of the cañon, later to be Gold Hill, to prospect some yellow dirt he had seen along the top of a low ridge or mound. The dirt was not rich, but when they began to work there with rockers in the spring they soon dug down to the top of what became known as the famous Old Red Ledge.

Thus the Comstock Lode was discovered, both on its north and south ends, in the spring of 1859 by two groups of poor placer miners, working a mile apart, who had no thought of finding ore.

The placer miners had long known the great vein (later called the Comstock Lode) which extended along the lower east face of a range of high, bare hills and crossed the heads of Gold and Six Mile Cañons.[5] But, like many other veins in that region, the bold ragged footwall croppings were practically barren. The rich ore bodies discovered later lay along the upper or hanging wall side of the lode and were covered with wash.

It is commonly stated that the erosion of the Comstock Lode at the head of Gold Cañon released the gold that fed the placer mines below, and that the miners followed rich gravel until it led to the discovery of ore at Gold Hill. Neither assumption is correct. It does not take nine years for miners to work up a little stream. Besides, they always prospect ahead to see if there is not better ground above. The early geologists, Baron Von Richthofen, Clarence King, and George F. Becker agree that the Lode was only slightly eroded after it became highly mineralized.[6] If erosion had been deep the placers below the Ophir would have been rich and extensive, and Gold Cañon would have been profitable above Devil's Gate instead of too poor to work.

The fact is that the gold came in large part from the disintegration of the many little gold veins that ribbed the hills in the vicinity of Silver City, where the side gulches leading into the cañon paid well for years, although much of the dirt had to be

---

[5]Those high hills, which came to be known as the Virginia Range, were favorite hunting grounds of the placer miners. There was always a little water in several of the ravines, and sagehens and rabbits were plentiful. Deer also were sometimes found, and, possibly, antelope on the flats.

[6]Becker speaks of the erosion as "trifling." *Geology of the Comstock Lode*, pp. 185, 271, 389, 390 (1882).

packed to water.[7]   These veins, worked in a small way as a rule by local owners, have continued to yield more or less gold from 1860 to the present time, but none proved rich until the Oest and Hayward mines were found about 1885.   Unfortunately, their rich ore lasted only a couple of years.

THE GROSH BROTHERS

The Grosh brothers might have discovered the Comstock if death had not struck twice in 1857, although that is not probable. The veins about the present Silver City which the brothers were prospecting showed some values on the surface and were much more inviting.   They have the honor of being the first in that region to prospect intelligently for silver, and there can be no doubt that their "monster ledge" was the Silver City branch of the Comstock Lode.   That large vein of low-grade ore was never profitable, although extensively developed by the Dayton, Kossuth, Daney and other companies.   Henry De Groot credited the brothers with discovering silver ore, but said:

It is obviously a mistake to say that they discovered the Comstock, unless we consider the ore channel that passes through the present Dayton claim (at Silver City) as being the main lode, for there can be no question that they obtained their best specimens from that neighborhood. * * * The fact is that they never found any ore like the rich silver ore at the Ophir.[8]

None of the old-time Comstockers believed that the brothers had discovered the Comstock.   Their prospecting led to nothing worth while and had no connection with the discoveries on the main lode nearly two years later.   The only profitable mines in the vicinity of Silver City were those located on little gold veins.

The struggles and the sad fate of the brothers have aroused the sympathy of thousands, including this writer, who regrets his inability to award them the honor of the discovery, notwithstanding the favorable views of Lord and Shinn among the authors cited.[9]

---

[7]Dr. V. P. Gianella states that the erosion was more extensive on those lower hills than along the Comstock. *Geology of the Silver City District* (1936). University of Nevada Bulletin, Vol. 30, No. 9.

[8]"Comstock Papers," *Mining and Scientific Press*, July 22, 1876.

[9]Dr. Richard M. Bucke's "Twenty Five Years Ago," *Overland Monthly* for June 1883; Dan DeQuille's *Big Bonanza*, pp. 33–35 (1876) ; Lord's *Comstock Mining and Miners*, pp. 24–32 ; Henry DeGroot in *Mining and Scientific Press*, July 22, 1876, p. 64 ; Report of Nevada Surveyor General for 1865 ; Shinn's

THE COMSTOCK DISCOVERED AT GOLD HILL

The little original Gold Hill mines, as previously stated, were located as placer claims. James Finney, familiarly known as "Old Virginny" (so called because of his native State), Alec Henderson, Jack Yount, and John Bishop made the locations on the rather poor dirt that covered a low flat-topped side hill, about 60 feet in height and nearly 500 feet long, which stood near the head of the right-hand fork of upper Gold Cañon.

"Old Virginny" had taken notice of this little yellow hill while on one of his hunting trips, as that color sometimes indicated the presence of gold in that region. The miners about Johntown, three miles below, were in winter quarters, waiting for the spring thaw to provide water to wash their nearly played-out gravel, and during a spell of good weather, on January 28, 1859, the four men went up to the hill to prospect it. There was snow on the ground, but they cleared some of it off and washed a few pans of dirt that they found at the mouth of a ground-squirrel's burrow. It yielded about fifteen cents to the pan, which was a fair prospect.

Each of them then located a placer claim 50 feet in length along the hill and extending 400 feet across the gulch, in accordance with the placer location rules of Gold Cañon. They called the new diggings Gold Hill to distinguish them from the placers down below.

A few days after the discovery, Lemuel S. "Sandy" Bowers, Joseph Plato, Henry Comstock, James Rogers, and William Knight came up from Johntown and located an adjoining 50-foot claim, which they later subdivided, each taking a 10-foot strip across the hill and the gulch.[10]

> The new diggings were discovered on Saturday and the next day (Sunday) nearly all of the male inhabitants of Johntown went up to the head of Gold Cañon to look at and "pass upon" the new mines. The majority of the sagacious citizens of the mining metropolis of the country (it contained about one dozen small houses of one

---

Story of the Mine, pp. 26–36 (1896) ; Thompson & West *History of Nevada*, pp. 50–54.

The Thompson & West *History of Nevada*, which will be referred to many times on these pages, is the most informative and useful history of Nevada up to the year 1881.

Shinn's *Story of the Mine* is one of the most readable and instructive books about the Comstock, notwithstanding frequent misstatements due to lack of personal familiarity and reliance upon undependable sources.

[10]*Virginia Daily Union*, October 9, 1863 ; Lord's *Comstock Mining and Miners*, pp. 35, 36 (1883).

kind and another) did not think much of the new strike.
They had placer mines near at home, three miles below,
that prospected much better.[11]

Nevertheless, other miners soon came up and located claims
along the hill.

In early spring, when the weather permitted, a number of the
miners began to wash the dirt through their rockers and were
delighted to find the ground richer as they dug down. Chunks of
decomposed quartz appeared in the soil, which they crushed with
picks and hammers to free the gold before putting the fragments
through their rockers. Sometime in March or April, at the depth
of about ten feet, the miners were astounded to dig into a vein of
broken reddish quartz fairly rich in gold, which became known
as the Old Red Ledge. They had discovered the Comstock Lode,[12]
although the ore carried little silver until the main north ore
bodies were encountered later. When their placer claims turned
into quartz claims they continued to hold them as placer.

The discovery at Gold Hill created a little local excitement and
the placer miners, ranchers, station-keepers, and others from miles
around came to locate claims both north and south of the hill.
There were about a thousand white people living in the region at
that time; all of them dependent upon the travel along the Old
Emigrant Road with the exception of the Gold Cañon miners.

The notices of location of the Comstock mines were the crudest
ever written, and the source of much litigation in later years.
They usually consisted of a line or two claiming so many feet
north or south from a stake or from another claim, with nothing
else to identify the location,[13] which made it easy at a later period
to "float" a claim over more desirable ground.[14] The confusion

---

[11]Dan DeQuille's *Big Bonanza*, p. 45 (1876).

[12]"Here it was that the Comstock Lode was first struck—though not the silver ore—by Old Virginia, John Bishop, and others." Dan DeQuille's *History of
the Comstock*, p. 99 (1889) ; Dan DeQuille's *Big Bonanza*, p. 46 (1876) ; Henry
DeGroot in *Mining and Scientific Press*, July 22, 1876. *The Nevada Directory*
for 1863, pp. 153, 306, says : "That mound-like mass of quartz rock was discovered by Finney and Kirby in March 1859."

The adjoining Yellow Jacket quartz mine was located May 1, 1859, shortly
after the discovery of the ledge at Gold Hill.

[13]The *Virginia Evening Chronicle* of August 30, 1878, printed many of those
old location notices under the caption "Relics of the Past." Other copies and
comments were printed by the *Gold Hill News* on April 10, 1880. Several of
the notices are printed in Lord's *Comstock Mining and Miners*, pp. 47, 48, 102
(1883) ; and others in Dan DeQuille's *Big Bonanza*, pp. 62, 63 (1876).

[14]"We shall never outgrow this perpetual litigation in mining matters until
the courts here shall rule that all indefinite floating claims and locations are
worthless." Correspondent of *San Francisco Bulletin*, July 14, 1862.

over titles was increased by the erasures and mutilations in the record book which was kept on a shelf back of the bar of a saloon. The Recorder, V. A. Houseworth, was a busy blacksmith.

The earnings of the Gold Hill placer miners soon began to increase to twenty dollars a day, but the outside world heard little about it until the discovery of the rich silver ore at the Ophir about two months later.

THE OPHIR DISCOVERY

"Old Virginny" and his friends thought they had finished the little placers at the head of Six Mile Cañon in the summer of 1858. They had worked almost up to the point where the Ophir discovery was made, but quit because the ground was poor and the clay troublesome. If they had dug three feet deeper they would have found the layer of rich black sand accidentally uncovered by O'Riley and McLaughlin the following June while digging out the little spring to increase the flow. The ground was poor and they were becoming discouraged when a test of the sand found in the spring covered the bottom of the rocker with pale yellow gold.[15] When they recovered from their astonishment they set to work feverishly, washing out more and more gold.

Toward evening, loud-mouthed, half-mad Henry T. P. Comstock happened along, riding a strayed pony, his long legs dragging the sagebrush. When he saw the gold he was off in an instant and exclaimed: "You have struck it boys." After running the glittering scales through his fingers and "hefting" it he straightened up and began to bullyrag the miners for taking gold out of the land that he had located for a ranch. Besides, he argued vociferously, he and his partner Emanual "Manny" Penrod owned the spring, and that the men who formerly worked there had some claim on that ground. Those old-time Gold Cañon miners seldom quarreled over ground as it usually played-out quickly, so O'Riley and McLaughlin agreed to take in Comstock and Penrod as equal partners.[16]

Penrod then joined them and he and Comstock bought out the men who had worked there in 1858 for $50, and Comstock gave an old blind horse for the one-tenth of the spring which he and Penrod did not own. The four then dug a wide trench up the

---

[15]The gold was worth only $8 or $9 an ounce, owing to the silver alloy.

[16]Penrod "who had an excellent reputation for honesty and good sense," gives a convincing account of the discovery and of subsequent events in Thompson & West *History of Nevada*, pp. 56, 57 (1881); Lord's *Comstock Mining and Miners*, pp. 37–39.

slope to uncover the six-inch layer of sand—"some of it black as soot," Penrod said, which yielded hundreds of dollars a day in gold. They cursed and threw aside the heavy bluish sand because it clogged their rockers, not knowing that it was rich silver sulphide. Then, about June 12, the pay dirt turned into a shattered, partly-decomposed mass of quartz and sand, stained with black

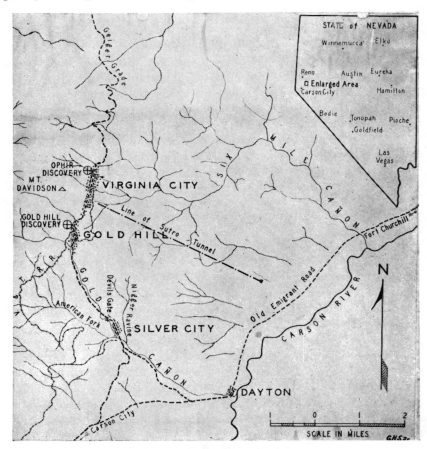

VIRGINIA CITY AND VICINITY

manganese and intermixed with heavy bluish-gray quartz. When this showed gold they pounded it up and washed the fragments through their rockers. They had never seen any of that "blue stuff" in the Gold Cañon placers, nor had any such ore ever been found in the mines in that vicinity. Dan DeQuille, who knew all of the early placer miners, wrote:

In mining in Gold Cañon they had been bothered with a superabundance of black sand and heavy pebbles of iron ore, but this new bluish sand was a thing which they had never encountered anywhere in the country.[17]

O'Riley and McLaughlin wanted to locate the ground as placers, 50 feet to each man according to the rules, but Penrod insisted that they had discovered a quartz vein and that it should be located as such, each of the four men taking 300 feet along the vein with an additional 300 feet for discovery. They measured off 1,500 feet with a piece of rope and set a stake at each end, which was all that was done to mark the boundaries. The Ophir notice of location was not recorded. Penrod states that a 100-foot strip was segregated from the Ophir claim, 200 feet from the south end, and given to him and Comstock for insisting upon locating the ground as a quartz claim and for the use of the spring. This 100 feet became the Mexican.

The slight value that the owners placed upon their mine is shown by the contract they made on June 22 with John D. Winters, Jr., and J. A. Osborn, whereby the latter were given a one-third interest in the Ophir for building two arrastras, worth $75 each, in which to work the ore, together with two horses or mules to operate them.[18]

Not until June 27, when some of the "blue stuff" was assayed by Melville Atwood at Grass Valley, California, was it known to be rich in silver. J. F. Stone, a station-keeper on the Carson River and former Grass Valley quartz miner, who located the Sierra Nevada claim, had sent the quartz over by a rancher, B. A. Harrison, who gave it to J. J. Ott of Nevada City to assay. His test showed it to be worth $840 to the ton. Harrison then gave a sample of the ore to Judge James Walsh, a prominent miner and mill owner at Grass Valley, who handed it to Melville Atwood to assay. The result of $3,876 to the ton, three fourths silver and one fourth gold, was so startling that Atwood assayed it a second time. Walsh sought the advice of Richard Killala, "a distinguished Irish metallurgist" then living at Grass Valley,

---

[17]Dan DeQuille's *Big Bonanza*, pp. 40, 48, 55 (1876). Dan went to Gold Cañon in 1860 and lived at Silver City for a year. Shinn was mislead in stating that the placer miners in Gold Cañon had been annoyed by "heavy blue stuff." He confused Gold Cañon with the Ophir Diggings. *The Story of the Mine*, p. 36 (1896). None of the early writers speak of "blue stuff" in the Gold Cañon placers, nor of any similar material.

[18]A copy of the contract is printed in Thompson & West *History of Nevada*, p. 57 (1881). It was recorded on June 22, 1859.

who encouraged him to go to Washoe.[19]   Meanwhile Stone had arrived and assured the Judge that the placer miners were throwing aside tons of that stuff.   Early the next morning the Judge and his partner, Joseph Woodworth, Stone, and Harrison left Grass Valley on horseback, followed almost immediately by alert miners.   The first "Washoe Rush" was on,[20] although a small affair compared with that of 1860.

The first newspaper mention of the discovery at Washoe was made in the "Nevada Journal" (published at Nevada City, California) on July 1, 1859, which stated that an assay of the ore by J. J. Ott showed it to contain $840 to the ton—one-third silver and two-thirds gold.   The editor, however, warned his readers:

Of course, the discoverers of this vein may have struck a good thing, but the odds are about 10 to 1 that instead of opening up their quartz vein they intend to open a provision store.[21]

The article told of the quiet departure of Judge James Walsh and his partner, Joseph Woodworth, for the new mines "after the manner of the miners in 1850."   The editor did not know of the rich Atwood assay.[22]

On July 2 the Placerville "Semi-Weekly Observer," edited by Prof. Frank Stewart, reported: "We saw a specimen of the Carson Valley gold quartz yesterday.   It has a bluish cast and looks more like common blue limestone than anything else.   The sample we examined was full of gold, however."   The placer miners who had made the discovery at the Ophir were saving only the gold until the Walsh party arrived with the news that the ore was rich in silver.   Hittell says that the reports "commanded little attention" until late in the fall when the first large shipments of rich ore reached San Francisco.   Californians had become skeptical about reports of new strikes—they had chased so many rainbows, beginning with the "Gold Lake Rush" in 1850.[23]

California was in the midst of a depression in 1859, resulting from the disastrous "rush" the year before to fabled gold placers

[19]Henry DeGroot in *Mining and Scientific Press* of August 12, 1876.   This is a corrected statement made at the request of Judge Walsh.

[20]Lord's *Comstock Mining and Miners*, pp. 35–40, 54, 55; Thompson & West *History of Nevada*, pp. 60, 61 (1881); Dan DeQuille's *Big Bonanza*, pp. 40–60 (1876); Henry DeGroot in *Mining and Scientific Press* of August 12, 1876.

[21]The *Nevada National* at Grass Valley republished the article on July 2, and it was reprinted by the *Sacramento Daily Union*.

[22]Almarin B. Paul had that assay certificate.

[23]*U. S. Mineral Resources* for 1866, pp. 17, 18, 24, 25, 26, tells of the various miner's rushes, beginning with that to Gold Lake in 1850.

on the Fraser River, B. C. One sixth of the voting population of the state,[24] 18,000 men, had sold or abandoned their mines and farms and all they possessed to make that long, hard trip by sea or land. Some gold was found on the bars, but no reward for that host of men, who straggled back, broken in spirit and in health to a disheartened people. No wonder the reports of rich silver mines in the deserts of Washoe were viewed with suspicion. Then "silver" worked a miracle—it instilled new life into California.

That discovery not only rejuvenated California but led prospectors to search for mines throughout all of the western Territories. Rich placers were found in Colorado in 1860, and in Idaho and Montana and Oregon in 1861. Quartz mining soon followed.

The discovery of silver on the Comstock was only second in importance to that of gold in California.

THE EARLY LOCATORS

We are indebted to Henry De Groot for the names of the locators of the principal Comstock mines, most of whom were Gold Cañon placer miners. A number of the leading mining companies adopted the names of the men that located the claims, for example: the Gould & Curry, the Hale & Norcross, the Best & Belcher, the Savage, the Chollar, the Overman, the Belcher, the Sides and the White & Murphy (which became the Con. Virginia). DeGroot lived in the cañon in 1859 and 1860 and knew them all then and later. Seventeen years after the discovery he wrote a series of articles for the Mining and Scientific Press[25] in which he outlined the lives and characters of those men and told of their small rewards. In one article he said:

> More than half of them are dead, those living are either in moderate circumstances or very poor; none rich. Most of them were brought to this pass by liberal and improvident habits. * * * A few of them were men of education and reticent, cautious, and thrifty; more, however, were garrulous, imprudent, and spendthrift, almost the whole of them being men of generous impulses and free-hearted.

---

[24]J. S. Hittell's *Mining in the Pacific States*, pp. 29–35 (1861).

Horace Greeley came across the Continent by stage in June 1859, and made his famous trip over the Sierras with Hank Monk early in July. Hank's admonition: "Keep your seat, Horace! I'll get you there on time if I kill every horse on the line," became one of the stock stories of the West.

[25]"Comstock Papers," *Mining and Scientific Press*, September to December 1876.

He comments on the susceptibility of those men to feminine wiles after their long exile in that region—"with disastrous consequence in many cases." They were not quite cut off from the outside world. The Old Emigrant Road to California ran along the Carson River, three miles below Johntown, and they obtained their supplies at Spafford Hall's station as well as some of the news of the day. In 1853 a small store was opened at Johntown by Walter Cosser.

Lord, a literary man and city dweller, has some dreary comments on the life of the placer miners.[26] He, of course, had no comprehension of that healthy outdoor life. Those men were free, their wants were few, their living assured. Time meant nothing to them—they slept, ate, and worked when they pleased. When there was water they worked hard; at other times they loafed. The hunting was good on the hills, the river teemed with trout. They owed the world nothing nor the world them. Dr. Bucke, in his story of the Grosh brothers, speaks kindly of those men as he knew them in 1857.

SNOW-SHOE THOMPSON

Our chief source of information concerning Gold Cañon miners is in news items furnished by John A. "Snowshoe" Thompson to the Sacramento Union, from January 1856, to March 1858; and from the excellent letters of "Tennessee" to the San Francisco Herald from 1857 to 1860, inclusive.[27]

It seems incredible that a lone man, carrying as much as 60 to 80 pounds of mail and express, should cross and recross the untrodden Sierras as many as thirty-one times in one winter, yet Thompson made such trips during four winters, beginning in January 1856, and as casually as if it were summer time. The first account of his method of travel appeared in the Sacramento Daily Union of April 22, 1857:

> Mr. Thompson, the Mountain Expressman, has crossed the Sierra Nevadas thirty-one times during the winter months, generally on snowshoes (skis) ; this trip, however, they were not necessary, the crust of the snow being strong enough to sustain him in his ordinary shoes. He never carries a blanket or other covering, save a common suit of winter clothes. When night overtakes him,

---

[26]Lord's *Comstock Mining and Miners*, pp. 19, 20 (1883).

[27]"Tennessee" somehow became marooned in the obscure hamlet of Genoa and consoled himself by writing letters to the *San Francisco Herald*. The Bancroft Library has a file. He wrote to the *Sacramento Union* occasionally.

he kindles a fire by some dry stump or tree top, and lies down by its side. The extraordinary exposures never produce colds, but as soon as he reaches the settlements and breathes the air of confined rooms he becomes subject to colds.

Thompson, a Norwegian farmer living on Putah Creek, California, heard that the settlers in Carson Valley and the placer miners at Gold Cañon were without mail and necessary small supplies in winter, and would pay well for such service. So, in the fall of 1856, he fashioned a pair of skis, such as were used in his native land, and became Santa Claus to the few people on the east side of the range. Those skis are in Sutter's Fort Museum at Sacramento.

Thompson was one of the volunteers in both battles with the Piutes at Pyramid Lake on May 12 and June 2, 1860. Myron Angel wrote of him:

> He was never lost in the woods or in the mountains. By observing the trees and rocks he could tell which was north and direct his course accordingly. He was a man of great physical strength and endurance, and of such fortitude of mind and spirit that he courted rather than feared the perils of the mountains.[28]

But Nature takes her toll; he died some years later near Genoa, Nevada, at the age of forty-nine.

"OLD VIRGINNY"

Picturesque, genial, open-handed "Old Virginny" has been the butt of many writers, chiefly because of his convivial habits, although Dan DeQuille says his sprees "were generally followed by seasons of great activity." When there was so much to his credit it seems a pity that he should be remembered only for his weaknesses. He was known as the best judge of placer ground in Gold Cañon; he located the first quartz claim on the Comstock, the worthless Virginia croppings (the footwall of the Ophir), on February 22, 1858; he is credited with the discovery of the placers below the Ophir in 1857; and it was "Old Virginny" who led the other three up the cañon on January 28, 1859, to locate the placers on Little Gold Hill. However, honor came to him: the early miners at Virginia City acknowledged their debt by

---

[28]Thompson & West *History of Nevada*, pp. 103, 104; Lord's *Comstock Mining and Miners*, p. 21. Dan DeQuille, who knew Thompson well, wrote a sketch of his life for Vol. 8 of the *Overland Monthly* (2d Series), pp. 419–435.

holding a meeting and naming the town after him,[29] and the mountains above the towns came to be known as the Virginia Range.

Money meant little to "Old Virginny"—generosity was second nature. He gave Little French John nine feet of his Gold Hill claim for "tending him through a spell of sickness," and received less for his Comstock interests than any other locator. Those who knew him best were not content to bury him without a tribute to his achievements and his character. After he was killed by being thrown from a horse, on June 20, 1861, the people of Dayton (then known as Nevada and as Mineral Rapids) gave him an appropriate funeral and passed the following resolution:

> Resolved, That in the death of James Finney, the people of this Territory have lost a man to whom more than any other they are indebted for the discovery of the mineral wealth of this Territory.

> Resolved, That while we humbly bow to this dispensation of Providence, we recognize it not less a matter of propriety than duty to give our testimony to the virtues of the deceased; and whilst acknowledging his faults—faults common to mankind—we deem it not complimentary, but just, to say that James Finney was ever known among the people of this Territory as a generous, charitable, and honest man.[30]

JOHN W. MACKAY

John W. Mackay, who became the Comstock's richest miner and most notable man, was among the first to reach the "Ophir Diggings," as it was then called. He had come from New York to California in a ship he had helped to build as an apprentice, and arrived at Downieville in 1851 at the age of twenty. Like many another young Argonaut, all that he gained in the years that followed was a wealth of experience that rounded out his native ability and his character.

In August 1859, when the newspapers confirmed the discovery of rich silver mines, Mackay and his "pardner," Jack O'Brien, bought a few supplies and packed their blankets over the mountains one hundred miles to the new camp. On the way over

---

[29]*Daily Territorial Enterprise*, September 24, 1859; *Sacramento Union*, October 28, 1859.

[30]*Sacramento Union*, July 8, 1861; reprinted from *Daily Territorial Enterprise*.

Mackay said: "All I want is $30,000; with that I can make my old mother comfortable." Fifteen years later, when his income from the Con. Virginia was $300,000 a month, Jack twitted him about that remark, to which Mackay responded with his slow smile: "W-well, I've ch-changed my mind." (Mackay stuttered painfully as a young man, but overcame it almost entirely, just as he did his lack of early education.)

When our two adventurers reached the ridge north of the Ophir and were ready to start down the hill, Jack said: "John, have you got any money?"

"Not a cent," replied John.

"Well, I've only got a half a dollar, and here it goes," said Jack as he threw that last fifty-cent piece far down the hillside into the sagebrush. "Now we'll walk into camp like gentlemen."[31]

THE OPHIR IN 1859

The method of mining was crude in the extreme. The ore body had been exposed by wide trenches for a length of three hundred feet through the Mexican, Ophir, and Central claims. Apparently all that the owners thought of was to disclose as much of the Lode as possible in the shortest space of time, and, what was more important, to extract the streaks of rich, bluish-gray silver sulphide ore for shipment to San Francisco. The ore body at that time was far from being the bonanza that was developed during the succeeding year. George D. Roberts, one of the early arrivals, reported that "the rich ore, worth from $3,000 to $4,000 a ton, is found in irregular veins from 1 to 4 inches in thickness, enclosed in a large vein from 10 to 15 feet wide. After the silver ore is sorted out the remainder is worked in arrastras and pays from $300 to $500 a ton."[32] Deidesheimer stated that on November 1, 1859, at the depth of 30 feet, the vein of silver sulphurets was

[31]R. V. "Dick" Dey, in *San Francisco Call*, July 21, 1902; William E. Sharon. Mackay confirmed the story to a little group of friends on his last visit to San Francisco in 1901. He added, "That's what Jack did with his money all his life. (H. L. Slosson.) Long after O'Brien's death, Mackay never spoke of him without emotion. He once said to E. C. Bradley, Vice President of the Postal Telegraph Company: "Jack was a great gentleman, and I loved him better than any man I have ever known." (He is not to be confused with Wm. S. O'Brien who was one of Mackay's partners in the Bonanza Firm.)

Dey was Mackay's confidential secretary for over thirty years and idolized him. The day that Mackay died in London (July 20, 1902), Dey dictated to a reporter for the *San Francisco Examiner* a review of Mackay's life and character which compels the deepest respect and admiration.

[32]*Sacramento Union*, October 5, 1859.

from 4 to 15 inches thick.[33] Maldonado began to smelt some
of his rich ore in little adobe furnaces, Mexican fashion,[34] soon
after he bought Penrod's one half of the Mexican.

The rude little settlement then called "Ophir Diggings" was to
grow into the metropolitan mining camp of Virginia City—so
named in honor of "Old Virginny" early in September 1859:

> The miners at Ophir Diggings have changed the name
> of that locality to Virginia City, in honor of Mr. Berry
> (should be Finney), one of the first discoverers of that
> mining locality, and one of the most successful prospec-
> tors in that region.[35]

Another similar item, giving the correct name, Finney, appeared
in the Sacramento Union of October 26.

The Ophir sent 38 tons of selected ore to San Francisco that
fall, which yielded $112,000 or almost $3,000 a ton; transportation and reduction charges amounted to $20,576, leaving a
net return of $91,424. It was packed on mules or shipped by
wagons over the mountains to Folsom, thence by rail to Sacramento, and from there to San Francisco by boat. The Central
shipped 20¾ tons that yielded $50,000 gross. This was excep-
tionally rich ore and contained substantial percentages of zinc
and lead and a little copper and antimony.

EARLY TRANSACTIONS

As soon as the vein was disclosed at the Ophir, other men
located claims, both north and south. The Central, with 150 feet
of the lode, adjoined the Ophir on the south and proved to contain a slice of the southern edge of the bonanza.

The first Californians to arrive were experienced quartz miners
from Grass Valley and Nevada City, and they soon bought the
majority of the interests in the Ophir and adjoining claims.
Others from the same region purchased nearly all of the little
claims at Gold Hill. A few San Franciscans acquired shares in
the Ophir, but most of the capitalists from the Bay region arrived
in 1860 and did not fare so well, with the exception of those that
bought into the Gould & Curry and the Savage.

---

[33]Lord's *Comstock Mining and Miners*, p. 61 (1883).

[34]"The Mexicans here, with their little adobe smelting furnaces, have con-
trived to extract from the same quality of ore more metal than the scientific
gentlemen in San Francisco say there is in it." Letter from Virginia City,
November 26, 1859.

[35]*Daily Territorial Enterprise* of September 24, 1859, reprinted in *Alta California* of September 28, 1859. (The *Daily Territorial Enterprise* was then published at Carson City.)

Judge James Walsh, in behalf of himself and his partner, Joseph Woodworth, bought Comstock's one sixth in the Ophir, and one half of the old California and other interests for $11,000,[36] and Penrod's one sixth of the Ophir for $5,500. This one third of the mine, together with some other smaller interests which they acquired, gave them the control. The Judge was a careful buyer. After obtaining an option from Comstock, he and Comstock took 3,151 pounds of ore to San Francisco, which sold readily for $1.50 a pound.[37]

J. A. Osborn sold his one-sixth interest to Donald Davidson and General Allen "for a good price," according to De Groot. Patrick O'Riley, who held out longest, received $40,000 for his one-sixth from J. O. Earl of San Francisco, who arrived late in the fall.

George Hearst, who was one of the first to arrive, wrote his friend Almarin B. Paul, "You will soon know of tons of silver and gold leaving here every month." He took an option on McLaughlin's one sixth for $3,000 and hurried back to Nevada City to raise the money. This was the beginning of the Hearst fortune.

Penrod sold his one half of the Mexican to Gabriel Maldonado, a Mexican of Spanish descent, for $3,000. Comstock received $5,500 for his half interest from Francis J. Hughes. The mine was known as the "Spanish" until Maldonado sold to Alsop & Co. in 1861 after a lawsuit over borrowed money.

Although later writers, unacquainted with conditions at the time, have scoffed at the sellers, Henry De Groot says that some of the prices paid, particularly the $40,000 to O'Riley, were thought extravagant. Lord remarks: "Weighing the chances of gain and loss as they stood in 1859 the prospectors had no cause to reproach themselves with lack of foresight."[38]

All of the writers tell of the later poverty and misfortunes of the original owners of the Ophir with the exception of Penrod and "the Winters boys."

Comstock's braggart tongue fastened his name on the lode although he was not one of the original discoverers at Gold Hill or at the Ophir. He was known as "Old Pancake" by his fellows because he never had time to bake bread. Dan DeQuille, who knew him well, said he was not quite right mentally, and wrote

---

[36]Comstock's agreement of sale to Walsh, dated August 12, 1859, is printed on pp. 72, 73 of Dan DeQuille's *Big Bonanza*.

[37]Lord's *Comstock Mining and Miners*, p. 61 (1883).

[38]Henry DeGroot wrote at some length of the sale in "Comstock Papers," *Mining and Scientific Press*, October 12, 1876. Dan DeQuille tells of some of the transactions in *Big Bonanza*, pp. 72–74 (1876). See also Lord's *Comstock Mining and Miners*, pp. 61, 62 (1883).

of his pretensions: "He made himself so conspicuous on every occasion that he soon came to be considered not only the discoverer but almost the father of the Lode."[39]

Judge Walsh and some of the other early purchasers sold their interests in the spring of 1860 to a number of San Francisco capitalists, who organized the Ophir Silver Mining Company with 16,800 shares, 12 to each of the 1,400 feet of the Lode held by the mine.[40] Walsh, who had a large interest, received a substantial fortune for those times. His partner, Joseph Woodworth,[41] and other owners joined the new company, among them the "Winters Boys," John D., Theodore, and Joseph, who came from Pleasant Valley, below Washoe, where their mother had a ranch. They acquired valuable interests along the lode and became very prosperous.

[39]Dan DeQuille's *Big Bonanza*, pp. 41–51, 55, 77–81 (1876).

[40]The number of "feet" in a mine was the length of the lode within its end lines. A deed to ten feet conveyed one tenth of a mine 100 feet long.

All transactions on the exchanges during the early years were in feet, regardless of the number of shares for which a mine was incorporated.

[41]Woodworth could have sold his Comstock interests for a fortune a few years later, but held on too long and left with a small one.

CHAPTER II

1860, A Year of Disillusionment—Deidesheimer Invents the Square Set—1861, A Craze for Mill Building—1862, New Discoveries—Renewed Confidence.

The Comstock should not be thought of as continuously populous and prosperous during the twenty years of its greatness, from 1860 to 1880. While it was the most important and productive mining camp in the world the fortunes of the towns and the population rose and fell with the prosperity of the mines. There were periods of bonanza and of borrasca, and others of fairly steady going. The stock market in San Francisco was responsive to conditions on the Comstock, which it helped to maintain at all times by keeping unproductive mines at work by means of the "assessment system." But for that the towns would have been smaller and less prosperous throughout their entire history.

Many of the thousand or more Californians who joined the first Washoe rush in the summer of 1859 soon went home in disgust. There was no placer ground worth having, they said, the best quartz mines were already located, and the barren region was "no place for a white man to live." Nearly all of the remaining adventurers had gone before the snow came, leaving at the most 300 men of various sorts to face a hard winter, with little more comfort than in an Indian camp—in dirty ragged tents, shacks of rough boards, tunnels and holes in the sidehills, and a few stone cabins. A correspondent wrote in November:

> The real mining population of Virginia City is very small, perhaps not one sixth of the number of inhabitants. The remainder are the prospecting gentry, speculators, loafers, and gamblers. * * * The town is full of gamblers and their booths are full every evening.[1]

The largest structure was a saloon and gambling hall, housed in a tent forty by twenty-five feet nailed over a board frame. There at least men could be comfortable. "We have had a dismal time of it," wrote a correspondent early in February, "but such are the hopes of men and the confidence in the mines that there has been but little complaint."

---

[1]*Sacramento Union*, November 9, 1859.
"About 100 miners at work in Virginia City." *San Francisco Bulletin*, October 26, 1859.

The winter of 1859–1860 was unusually long and severe. A foot of snow fell on the 4th of November, followed by other storms that covered the ground to a depth of three to four feet. All outside work on the mines was suspended; the only work going on was the driving of tunnels to tap the Lode at shallow depths. There was much open weather, however, as is usual there, and men continued to locate mining claims over the snow without regard to what might be underneath. The mail brought news that the Californians were in a fever of excitement over Washoe mines. Apparently all wanted to buy claims or feet in the new silver mines without knowing anything about the location or the value, and the local prospectors were preparing to supply the demand. Owing to the scarcity of money during the winter, there was much trading and gambling in interests in mines.

The region for miles around was ribbed with quartz veins, few of which had any value whatever. All were soon located and many of them sold to silver-mad investors. Claims of that kind were soon abandoned, but with every new excitement all would be located again. First and last over 16,000 claims were located in the Comstock region.

The rock formations of those Nevada mountains, which were nearly all of igneous origin, were so unlike that of California, and the veins were of such a different character, that the early prospectors had little to guide them in making locations. A panning test revealed only the gold and sulphurets. They soon learned, however, to make a simple test for silver by heating the sulphurets, together with acid, in a flask and precipitating with salt solution.[2]

In February the population began to increase. The sixty tons of rich ore from the Ophir and the Central that reached San Francisco late in the fall dispelled all doubts as to the riches of Washoe and created unparalleled excitement throughout California. Thousands were waiting impatiently to cross the Sierras in early spring, chiefly over the Placerville route. This was a new and poor road leading up the American River to Johnson's Pass, built to supplant the old Emigrant Trail of gold rush days which ran along the ridges south of the river. The famous highway to be built a few years later was then a dream.[3]

---

[2] Dan DeQuille's *Big Bonanza*, pp. 111, 112 (1876).

[3] The road at that time did not follow the river to Strawberry Station; but, to add to the perils and difficulties of the pilgrims, it crossed the river on Brockless Bridge at the entrance to the canyon, and turned steeply upward to

The snow was not deep on the Sierras that winter—not exceeding ten feet—and the road was passable most of the time. The
more hardy adventurers began to cross over the packed snow in
pleasant February weather on foot or on horses or mules. In
March the number swelled to a thousand or more, and in April,
when the road was lined with traffic, a storm set in that blocked
all progress for a few days. Men with animals hastened back to
the lower regions, as there was no feed left. Sutro in the Alta
of April 14 described it as "a second flight from Moscow." Other
storms followed intermittently. The road was a morass of snow
and mud and the track was lined with disaster. Broken wagons
and merchandise littered both sides. Loaded pack mules struggled frantically to make headway in a line of wagons, horsemen,
burros, and men on foot carrying their blankets and perhaps pick
and shovel. It was a mad mob with but one thought—to get to
Washoe ahead of the others and have first chance at its riches.
A more motley crew never went on a miners' rush.

The mad rush over the Sierras was checked during May and
June by the childish panic which accompanied the ill-advised
Piute Indian War, in which fifty men lost their lives.

The Indian War completed the demoralization that set in as
soon as it became apparent that the tales of the wealth of Washoe
were false; that the only rich mines in the region were those
located on the ore bodies at the Ophir and Gold Hill; and that
those mines were so slightly developed that their future was
uncertain. The news spread quickly and the San Franciscans
who had clamored to buy interests in mines without knowing
anything of their location or value were now eager to sell—with
no takers. The market for mines fell flat and the outside public
lost interest.

Work on the mines was practically suspended during the summer of 1860, except on the Ophir and Gold Hill ore bodies. As the
year advanced the ore body at the Ophir developed into a bonanza
on the 160 foot level, and extended into the Mexican. The Gould

the top of Peavine Ridge, which extends along north of the river. The road
wound along the top of the ridge for about twelve miles before turning down
sharply to the river, with an easy grade of four miles to Strawberry.

There are many interesting and informative letters in the San Francisco
newspapers giving details of trips over the Placerville road and conditions at
the mines, from February to June 1860; more particularly in *San Francisco
Bulletin*, March 3, 5, 24, 30, April 9, 11, 21, 23, 25, May 9, 11, 12, 15, 17, 18, 24,
25, 26; *Alta California*, April 4, 11, 19; May 11, 13, 15, 16, 17, 25; June 1;
Golden Era, April 15; *Sacramento Union*, April 28, May 5–8, 26, 1860. The
May letters tell of the Piute Indian War.

SQUARE SET MINING

& Curry found a small vein of good ore near the surface which later proved to be the top of a great bonanza.

The little mines at Gold Hill were quarrying rich decomposed gold ore from a vein ten to twelve feet wide along the top of the small hill; the more friable ore was washed through rockers, the remainder reduced in arrastras and in the mills of Paul and of Coover & Harris which began operations in August. "Sandy" Bowers and Joe Plato found a concentration of soft, rich gold ore a few feet below the surface, yielding $30,000, which started "Sandy" on his lavish spending career. Plato on the other hand was thrifty. His claim was equally rich for its size but little was heard of it.

Perhaps ten thousand Californians of all sorts and conditions came to the Comstock in 1860, half of whom drifted back denouncing Washoe as a humbug. Less than four thousand remained in Virginia City, Gold Hill, and vicinity.[4] Hundreds of adventurous prospectors scattered far and wide over regions never before trodden by white men in search of other Comstocks. Aurora was discovered that year and some other promising mining camps.

Little was accomplished that summer except the building of the towns of Virginia City,[5] Gold Hill, and Silver City, and the construction of a few mills. Before winter the people of Virginia City were fairly well housed in 868 dwellings.

While the majority of the people were fairly decent and law-abiding, the camps were infested with low-class gamblers, saloon roughs, pimps, and other human vermin that preyed upon the unsuspecting and quarreled and killed among themselves. There was only a shadow of law enforcement, and the six violent deaths in Virginia City in 1860 were almost unnoticed by the better class—except to wish there had been more of them.

Deidesheimer Invents the Square Set

The important event of the year 1860 to the mining world was the design of the square-set system of timbering by Philipp

---

[4]The U. S. Census taken in August 1860 gave Virginia City a population of 2,345, of whom 139 were females; two thirds were native Americans, the remainder made up of Irish, Germans, English, and all other nations. There were 42 saloons, 42 general stores, 9 doctors, 1 dentist, 2 school teachers, 1 milliner, but no preacher and no lawyer. There were five lawyers at Carson City, however, where Wm. M. Stewart was prosecuting attorney.

[5]Virginia City, October 23, 1860. Winter near at hand and freights over the Sierras 10 to 12 cents per pound; business fair; wages: mechanics, etc., $8, carpenters $6, laborers $4, cooks $100 per month. Methodist and Catholic churches being built. (*San Francisco Bulletin*, October 27, 1860).

Deidesheimer, a young German engineer and graduate of Freiberg, who had been mining at Georgetown, California.

The Ophir ore body, which had been only 10 to 12 feet wide on the 50-foot level in April of 1860,[6] had increased steadily in size until in October it was from 40 to 50 feet in width, in places, on the 180-foot level, and so soft and unstable that no known method of timbering would permit its extraction. The ordinary cap and post were of no avail, for even if made long enough they could not withstand the weight. Pillars of ore were out of the question. In this strait, W. F. Babcock of San Francisco, one of the trustees of the mine, sent for Deidesheimer, who arrived in November and worked out the plan.[7]

Comstock ore commonly occurred in large thick bodies. The quartz was not only crushed and water-soaked, but rendered still more unstable by the presence of clay and inclosed fragments of porphyry. The surrounding walls were of clay, which were only a few feet in thickness as a rule, but occasionally increased to 20 or 30 feet. Beyond the clay were belts of partly decomposed porphyry and other walls of clay, with occasional parallel sheets of ore. When an opening was made the whole country began to swell and move unless held back by stout timbering. The expansive power of the clay almost passes belief:

> It is of tough consistency, and when air is admitted by gallery or shaft it immediately begins to swell and exert tremendous pressure, forcing itself through the interstices of rocks, bending and breaking the most carefully laid timbers and filling mine openings with extraordinary rapidity.[8]

1861, A CRAZE FOR MILL BUILDING

Some progress was made during 1861.[9] President Lincoln signed the bill creating Nevada as a Territory on March 2, 1861. He then appointed his friend James W. Nye as Governor. The

---

[6]*San Francisco Bulletin*, April 23, 1860.

[7]Dan DeQuille's *Big Bonanza*, pp. 134–136 (1876). Dan wrote a more detailed description of the invention with illustrations for Thompson & West *History of Nevada*, pp. 572–574, to which Deidesheimer attached his approval.

[8]Clarence King, who wrote the preliminary geological report in Hague's *U. S. Exploration of 40th Parallel*, Vol. 3, p. 50. See account on p. 61 of a 10 x 12 chamber opened in clay in the Savage to obtain stope-filling, which filled as fast as men could wheel it out for several weeks, and then closed up.

[9]The population of Virginia City nearly stood still until 1863. A second census taken in August 1861 gave it only 2,704 resident inhabitants. In the fall of 1862 the number was estimated at a little less than 4,000. Meantime the population of Gold Hill and Silver City had doubled from 1,250 to 2,500.

people settled down to more normal living and went forward with improvements and mine development. The Ophir shaft reached the 300-foot level where the bonanza was still wide and rich. Its big mill at Washoe, on Washoe Lake, was approaching completion. The Gould & Curry found more ore, although nothing startling until the end of the year. The Potosi discovered a fine body of ore not far below the surface near the east wall, and was soon bitterly litigating with its parallel neighbor, the Chollar. The twenty Little Gold Hill mines were richly rewarding the many owners, who were spending their money freely. Aurora (Esmeralda) became the center of excitement in Nevada that year. The numerous rich veins that cropped on the surface and the boulders of rich ore scattered about led men to believe that the camp would eclipse the Comstock.

The year 1861 was marked by a craze for building mills. Not only were mills erected by the mines that had ore, but by others that expected to find it; many were built to serve as custom mills; some of the latter were small second-hand affairs brought over from California. A correspondent of the Alta California[10] names seventy-six mills built to handle Comstock ore—some in Virginia City and Six Mile Cañon below; many others in Gold Cañon from Gold Hill to the Carson River; a dozen along the Carson River from Empire to Dayton, and seven more in Washoe Valley. The number of mills is imposing, but there were only 750 stamps in all, and no ore for half of them.

The following winter was the most dismal and disastrous ever endured on the Comstock. A heavy fall of snow in November was followed by widespread and unprecedented rains in December, which melted the snow and filled all the ravines and cañons with torrents of water. Houses and mills were washed away or demolished and roads were torn out; the Carson River raged, carrying away bridges, mills, and habitations. Several people were drowned. The Ophir and adjoining mines, with their large open surface cuts, were nearly drowned out. The Little Gold Hill mines, which were open to the sky, filled with water and caved, and so remained for six months.

1862, New Discoveries—Renewed Confidence

An early spring in 1862 brought renewed confidence. Men set about repairing the damage and by midsummer hopes were high. The year exceeded all expectations. The big Ophir mill went into

---

[10]Letter printed in the issue of April 8, 1862. The *Nevada Directories* of 1862 and 1863 enumerate and describe most of those mills.

operation on high-grade selected ore; the Gould & Curry was engaged in the construction of its monumental mill in Six Mile Cañon; other mines and mills resumed. Stocks began to climb, stimulated by local developments and by glowing reports from Aurora, Austin, and the Humboldt region.

When further development work in the "D" Street tunnel of the Gould & Curry disclosed the largest and richest body of ore found on the Comstock up to that time, and the Savage encountered its extension, expectation knew no bounds. Stocks rose spectacularly; Ophir from $1,225 a foot in April to $1,500 in July, and to $3,800 in October; Gould & Curry from $500 to $1,050 and to $2,500 during the same period. The few other Comstocks then quoted on the market rose with the leaders. Aurora stocks, headed by the Wide West, were advancing riotously.

The rapidly increasing demand for stocks led to the organization of the San Francisco Stock and Exchange Board on September 11, 1862, the first mining exchange in the United States. Up to that time such transactions had been limited and were handled by the few brokers in their offices or on the streets. The Exchange focused public attention on the stock market and heightened the speculative spirit. By the end of the year 1862 the "Boom of 1863" was well on its way. Warnings were wasted. A level-headed correspondent in Virginia City wrote on October 12, 1862:

> Are all the folks at the Bay stock mad? Are you all so rich that you can recklessly create a mania for Washoe stocks? Go it blind, or what is worse, be blindly led into speculations, which will end in ruin, by clever chaps who know a "hawk from a handsaw" * * *. But I see you have a board of brokers, too. Maybe that accounts for the milk in the cocoanut. * * * Barnum was right. * * * Have the people become insane that they should unreflectingly rush into a financial abyss?

That was almost a lone voice. The Carson Silver Age of October 5 announced in a burst of enthusiasm: "People are astounded at the fabulous riches of the mines. * * * Within three years the mines of the Territory will produce $150,000,000, and the population will reach 100,000." The fabulous riches of which the editor boasts were those disclosed in the Ophir, the Gould & Curry, and the Little Gold Hill bonanzas.

The other mines in the Gold Hill section which afterward became famous—the Yellow Jacket, Kentuck, Crown Point, and the Belcher—were only promising prospects in 1862. The Yellow Jacket and the Belcher discovered their first ore bodies in 1863; the Crown Point in 1864, and the Kentuck not until the end of 1865. Nevertheless, a rush to Nevada followed in 1863 which dwarfed that of 1860 and established the Comstock as a great mining camp.

Reports of the riches of Nevada mines stimulated that great and daring enterprise, the Central Pacific Railroad, which was initiated by ground-breaking ceremonies on January 8, 1863, at Sacramento. Construction of the road was commenced at once and carried forward with energy.

The production of the Comstock increased slowly during the early years. In 1859 it did not exceed $275,000; in 1860, the year of the first boom, the total yield was only $1,000,000; during 1861 it rose to $2,500,000, and in 1862 to $6,000,000. The first year of the large production was 1863, when the riches of the Ophir and the Gould & Curry brought it up to $12,400,000.

CHAPTER III

The Boom of 1863—An Orgy of Litigation—Social Conditions—
The Placerville Road — Geiger Grade — Early Prominent
Men.

Never again was the Comstock to witness such turbulence, such feverish excitement, and such almost childlike anticipation as during the year 1863. It was a year of great growth in population and building, of gambling in stocks, and of an orgy of litigation over mining claims and mining rights. The rush over the mountains began in the fall of 1862, slackened somewhat during the open winter, and became a torrent in the spring. The old Placerville and Henness Pass roads (now rebuilt into great highways) were jammed with traffic; the other roads and trails carried an eager throng. Fully 25,000 people poured into Nevada, the greater number to the Comstock, the others to Aurora and Austin and to the many little camps that had sprung up throughout the Territory. The mining camps of California were stripped of their most vigorous and useful citizens; not only of miners but professional and business men of all kinds and classes. The cities too sent their quota. The scum and the dregs came also.

Social Conditions

Virginia City increased in numbers from 4,000 in the fall of 1862 to 15,000 or more by midsummer. Considerable progress in building had been made during the two preceding years, but now it came with a rush. Virginia City in one season became quite metropolitan. Homes and business buildings and office blocks were built, many of them of brick; gas and sewer pipes were laid in the principal streets; the International Hotel was enlarged to a hundred well-furnished rooms—with an elevator from the "C" Street entrance;[1] the daily stages brought in all of the luxuries of the Bay region; the restaurants vied with those of San Francisco; three theaters played to crowded houses. Maguire's Opera House,[2] seating 1,600 and elaborately equipped,

---

[1] The International was built in 1860 of one story and basement. In 1862 it was rebuilt of brick and much enlarged. (*Nevada Directory* of 1862, p. 137) : the hotel of 1863 was destroyed by fire on October 26, 1875, and the last and greatest hotel was erected in 1876. This was burned in 1916.

[2] John Piper bought the theatre of Maguire in 1867 and it became known as Piper's Opera House. Maguire's was located on the east side of "D" Street, near Union.

was opened on July 2 by young and talented Julia Dean Hayne, assisted by a distinguished company which remained a month, presenting Shakespearean and other plays; "Miss Lotta" Crabtree (then 16 and soon to become famous)[3] came in July and entertained the Comstock for three months with her joyous singing and dancing and drollery, which invariably brought showers of gold and silver coins. Two fine churches were built, making four in all; Sunday schools and public and private schools flourished; three excellent daily newspapers, together with those brought in by the daily stages, supplied a vigorous reading public. Many visitors from San Francisco were to be seen on the streets, who added not a little to the gaiety of the life; the "high-toned" saloons and gambling houses were elaborately furnished—the others fitted out according to the class of their customers; each race and kind had its favorite saloons. There were but two cardinal sins on the Comstock—lying and cheating at cards.

To a stranger—who sees only the obvious—the principal activities of the town appeared to be the stock market, drinking, gambling, and loose living. But that was misleading; mining camps always wear their worst side out.[4] The decent people far outnumbered the others. The town was made up of two classes, each of which kept its own place and went its way without interference from the other, unless the lower element came into collision with the law, which was not too strictly enforced. "Every man has a right to go to hell in his own way" was a common expression. Conduct was largely a matter of personal choice and more admirable than might be expected.

It was a period of heavy drinking and reckless living on the part of many; of lavish expenditure and equally lavish generosity and benevolence. The men about town prided themselves on being able to drink like gentlemen. They smoked Havana cigars and drank their whiskey straight—perhaps because you can take it oftener that way. Cigarettes were the despised badge of Mexicans, and "the girls" on "D" Street. They were a robust,

---

[3] The story of Lotta's life is well told in *Troupers of the Gold Coast*, by Constance Rourke, who somehow failed to learn of that Comstock engagement.

[4] Thomas Starr King, the famous San Francisco clergyman and orator who proved a tower of strength to the Union cause in California, received the usual unfavorable impression of Virginia City when he came there to lecture in 1863. Later, he described it as "A city of Ophir holes, gopher holes and loafer holes."

"Within the city limits there are some of the most magnificent palatial saloons to be found on the Pacific Coast, and not a few mean, dirty, low dens that would disgrace the Five Points of New York." (*Virginia Evening Bulletin*, May 3, 1864).

free-living lot; such people as Shakespeare wrote for, and their love of his plays arose from understanding hearts.[5]

Practically all of the men who came to rule the mines, the business, and the politics of Nevada had been youthful, adventurous, romantic California pioneers, and were in the prime of life; men of exceptional ability and resourcefulness, tried by hardships and ripened by experience. Judge C. C. Goodwin, who was a pioneer in both States, wrote in later life: "California drew to her golden shores the pick of the world, Nevada drew to herself the pick of California." These men centered on the Comstock where the rewards were greatest, although they scattered far and near to lead in the affairs of Nevada. While accomplishing wonders in that desert region they never seemed to take life seriously. Good fellowship was the prevailing spirit and everyone took part in whatever was going on—whether a parade, a picnic, a ball, a theatrical performance, a horse race, a prizefight, or a funeral. Life was a great adventure. Somehow they gave the impression of never reaching settled middle age, then or later. Humor was the spice of life. "Virginia is surely a pleasure-loving city" wrote the editor of the Gold Hill News. The Comstock celebrated everything.

The leaders dressed well—in black broadcloth as a rule—and bore themselves with quiet dignity. Their wives, who had shared the life in California, were worthy helpmates and held by the men in chivalrous regard. They were the homemakers, although taking part in all good works. It was a man's world. The newspapers and the stage were clean to the point of delicacy, out of respect for the ladies; suggestive jokes and even mild profanity were taboo.[6] Those rules, however, did not govern the men among themselves. Their talk was free and profanity a staple of conversation, as in old California days.

---

[5] All of the leading theatrical companies that came to the Comstock for years included Shakespeare's plays in their repertoire.

[6] The newspapers carried their gallantry so far that they refrained from publishing the names of the ladies except to note marriages, births, and deaths. No mention was made of social affairs. One would not know that the ladies attended the many balls and other entertainments given during the winter, excepting those given in connection with the churches.

Mistress Minerva Morris, in a letter to the *Gold Hill News* of October 13, 1863, complained that gallantry was carried too far and asked that credit be given for the refining influence of the ladies. She defends social conditions vigorously; "wives, maidens, and children abound, husbands are about as good as in California, and society and fashion flourish." A correspondent of *The Golden Era*, fully confirms Mistress Morris. *Golden Era*, May 22, 1864. Courts heard divorce cases behind closed doors.

The Comstock was Victorian in dress and manners and in its reading. To be a lady or a gentleman was good form; even the lower orders felt that influence.

No people were ever more friendly or more interested in one another. They were adventurers in a remote and barren region, mutually dependent in sickness and in health. Many of them had been friends in California mining camps and had many memories in common. The ties that bound them together pass the understanding of modern city people. The most striking characteristic was the independence and the individuality of the people. The notable men and women were deeply respected in an independent sort of way—for never was the rank and file more self-respecting. Ordinary men became transformed in that atmosphere, taking upon themselves some of the qualities of the greater spirits.

There never was another mining camp like the Comstock. The California pioneers who dominated it during the first twenty years and infused it with their spirit could live but once. Arthur McEwen, whose brilliant newspaper life began in Virginia City, wrote in later years: "The life of the Comstock in the old days never has been written so that those who did not share it can understand; it never can be so written.[7]

Virginia City had a lure for cultured and romantic men and women who sought fortune and escape from the monotony and conventionality of ordinary social life. The editor of the Evening Bulletin, under the caption "Talent at a Discount," deplores the fact that "Virginia is nearly overrun at present with editors, artists, and scientific men out of luck."

Meantime, the Comstock was a fools' paradise. Everyone had stocks in a dozen different mines—mostly wildcats—that were sure to make him rich. All of the races of the earth were there and all getting a thrill out of life such as few of them ever experienced again. "Money was as plentiful as sagebrush on the hills." The only money in circulation in the West was gold and silver, although in the East—owing to the exigencies of the Civil War— nothing but paper money was in use. Hardly anything could be bought on the Comstock for less than a quarter—twenty-five cents.

The Ophir, Gould & Curry, Savage, and the Little Gold Hill mines were turning out streams of bullion and their bonanzas were confidently expected to last for years; the Mexican was at its best until July 15 when the mine caved from top to bottom, partly wrecking the Ophir workings. The Chollar and the Potosi

---

[7]*San Francisco Examiner*, January 22, 1893.

were doing well, despite scandalous litigation. Great enthusiasm was aroused by the discovery of a rich ore body in the Yellow Jacket at the depth of 180 feet, and by another found in the Belcher in a shallow tunnel. The Sierra Nevada boasted of a mass of low-grade gold ore that gave promise of greater things.

Now the known ore bodies extended for a length of two miles along the Lode. Why not for five or six miles? The people of Virginia City and Gold Hill, which were built over the Lode, fancied themselves walking on streets of silver and gold, which was literally true while the ore bodies lasted. A local editor jubilantly proclaimed: "Every hill and cañon for miles around seems to be literally made of silver—there is no end to it." Another writer asserted that "The Comstock is a true fissure vein, the bottom of which can never be reached by man, and its great mineral wealth can never be exhausted." The yield of the mines increased from $6,000,000 in 1862 to $12,400,000 in 1863.

Over 400 new mining companies were operating (or pretending to operate) in the district, whose stocks were freely bought and sold on the three local exchanges at high prices. Daily reports of rich strikes in those mines were featured in the newspapers, for which compliant reporters received stocks. The shares of the principal mines were listed on the San Francisco Exchange from which daily reports came by telegraph.

Stocks rose rapidly all during the spring and early summer; shares bought today could be sold tomorrow at a profit. The stock market reached its peak at the end of June when Gould & Curry sold for $6,300 a foot, Savage for $3,500, Ophir for $2,100,[8] Belcher for $1,500, Yellow Jacket for $1,200, and all of the other mines along the Lode at proportionately high figures.

Those prices for leading stocks were not extravagant, considering the amount of ore in sight in the mines, and the fact that stocks were then selling by "the foot" regardless of the number of shares for which the mine was incorporated. The basis was the number of linear feet along the Lode owned by each mine, for example:

Ophir, with 1,400 feet, had 16,800 shares, or 12 to the foot; Gould & Curry, with 1,200 feet, had 4,800 shares, or 4 to the foot; Savage, with 800 feet, had 800 shares, or 1 to the foot; Yellow Jacket, with 1,200 feet, had 1,200 shares, or 1 to the foot.

---

[8]Ophir, which sold for $3,800 six months earlier had already suffered a sharp decline, owing to diminishing ore on the lower levels, floods of water, and the bitter and costly litigation with the Burning Moscow.

After 1865 all stocks were sold by shares which were largely increased.

People were speculating on hope. Out of all those mines, only three—the Ophir, the Gould & Curry, and the Savage—were paying dividends, and nearly all of the others levying assessments. The Little Gold Hill mines were very prosperous, but being private enterprises no information was published of production or of dividends, until the Empire and the Imperial began to pay dividends.

The total market value of all Comstocks at the height of the market was $40,000,000. At the same time the total assessed value of all of the real and personal property in San Francisco, where most of the stocks were held, was only $80,000,000. It was a mad West.

The Alta California, leading newspaper in San Francisco in the 50's and 60's, tells, half humorously, of that city's dependence upon the Comstock in 1863:

> So far as direct ownership and investment go, San Francisco is interested far more in the silver mines of Washoe than in the gold mines of California. The people of this city own ten times more stock in silver mines than in gold mines. * * * Not only the capitalist and the banker, but the wholesale and the retail merchant, the lawyer and the physician, the preacher and the editor, the carpenter and the blacksmith, the jeweler and the hotel keeper, the tailor and the shoemaker, the drayman and the stevedore, the clerk and the laborer, the milliner, the teacher, the manservant, and the maidservant—all own shares in some silver mine. It is a rare thing to meet a man who has not a certificate of stock in his pocket. Such a universal investment never was seen.
>
> Two thousand mining companies have been incorporated in this city within three years—one company for ten men. Wherever ten men are collected together in San Francisco, there is an incorporated mining company amongst them, each man owning, on an average, stock to the nominal value of $20,000. Whittington went to London with the expectation of seeing streets paved with gold; if he had lived in California he might have come to our metropolis expecting to see streets paved with mining certificates.

Peace, the reign of love and good will, the millenium—all these will be welcome in San Francisco, but the era for which we especially long is that when our mining stocks will be worth par in the market. Then we shall be worth about $500,000,000 more than we are now.

It is a strange turn in the career of fortune that the chief city of the land of gold has turned its back upon the auriferous deposits of the Sacramento basin, to run after the silver lodes beyond the mountains.[9]

Stocks began to fall slightly in July, followed by a slow and constant decline to the end of the year. There was no sharp crash such as followed the later booms of 1872, 1875, and 1878. People were buying stocks as a speculation and held on bravely. They had not learned to gamble on close margins and sell wildly on a declining market and bring on a panic.

A shock was given the insiders at the end of June 1863 when the Gould & Curry lower tunnel, 2,000 feet in length, hit the ore body. Instead of disclosing the deeper extension of the bonanza, it crossed 15 feet of low grade ore containing some streaks of high grade. This was 425 feet vertically below "A" Street where the tip of the bonanza was found two years earlier.

They hoped, and expected, that the rich ore would come in below, but came to learn the sad truth that the bonanza had pitched across the line into the Savage. Some fair to low grade ore was mined 50 feet deeper, and then the end. The Ophir, from which so much had been hoped, was showing signs of failing below the 300-foot level. Its last big year was in 1863 when 15,000 tons yielded $100.04 a ton. The first of a long record of assessments was levied in 1865. The theory of the inexhaustibility of silver mines was rudely shattered.

Unwelcome information was withheld from the general public as far as possible, and the Gould & Curry continued to give confidence. It had large reserves of excellent ore above the tunnel level and did not cease to pay dividends until three years later.

Some of the wise insiders, who were fully acquainted with the unfavorable developments of the lower levels of the Gould & Curry and the Ophir, and with the unsatisfactory performance of their great mills, began to sell their stocks and cleaned up fortunes. Thousands of others thought the decline was temporary and held on until all was lost. It was unthinkable that the Ophir, the Gould & Curry, and the Little Gold Hill bonanzas,

---

[9] *Alta California*, October 29, 1863.

and all of the other fine ore bodies along the Lode, would "play out" at an average depth of 500 feet, and that only fair ore bodies would be found thereafter to keep the camp going until the discovery of the great Crown Point and Con. Virginia bonanzas in the early '70s. Not even the wisest miner could have anticipated that.

Periods of great excitement in mining camps invariably attract large numbers of the depraved and the vicious, and crimes of all descriptions multiply. 1863 was such a year in Virginia City.

The town was almost free from violence until midsummer, but that was the lull before the storm. During the latter half of the year, accounts of gun fights, robberies, burglaries, brawls, petty thieveries, and drunkenness filled the columns of the newspapers, shocking and alarming the community.[10] The influx of depraved characters was so great that the police were unable to cope with them. Two policemen were killed and one seriously wounded while attempting to arrest drunken men. Two men were killed by women of the redlight district.

In a riot between rival volunteer fire companies, after they had quelled the disastrous fire of August 28, 1863, one man was killed and his assailant sentenced to two and one-half years. Thirteen homicides of all kinds were committed—equal to the total number during the four preceding years—and ten of the thirteen were the result of saloon brawls, in which most of the would-be bad men were eliminated. It was a good riddance, but makes a bad record. During the four succeeding years there were only thirteen homicides from all causes, after which such affrays constantly declined. From 1870 to 1880, when the population was greatest, there were but fourteen homicides from all causes in Virginia City.[11]

The bloodshed of the Civil War appeared to stir men's minds to violence throughout the West. The California newspapers recorded an unusual number of fatalities.[12] No deaths resulted

---

[10]Proximity to California is not considered from a very flattering point of view by the *Virginia Evening Bulletin*. That paper says Washoe cities and villages are flooded with the most desperate set of wretches that ever disgraced humanity. Persons escaping from San Quentin make a bee line for Silverado the moment their legs are free from fetters. Dishonest tradesmen, runaway wives, played-out gamblers, and the votaries and disciples of every vice—experts from California not indigenous to the soil—flourish vigorously on the Eastern slope. (*Golden Era*, November 1863).

[11]Thompson & West *History of Nevada*, pp. 341–355, lists the Nevada homicides from 1859 to 1881.

[12]Crime in California—Either something in the habits and customs of our people, or the times, or the atmosphere of the season seems to stimulate an unusual amount of violence, crime, and bloodshed. (*San Francisco Bulletin*, February 17, 1864).

EARLY-DAY TRANSPORTATION   (See bottom opposite page)

from Civil War friction on the Comstock. Such disturbances were chiefly vocal. Nor were any men killed in the many early conflicts between rival mining companies, in which armed men were employed. The lawyers for the companies were given credit for that restraint.

Virginia City never harbored a gang of professional killers, like Aurora, Bodie, Tombstone, and some midwestern towns; nor was the region terrorized by gangs of outlaws. The few stage robberies in the vicinity have been multiplied like Falstaff's men in buckram. Every stage robbery between Placerville and Virginia City was connected with the Comstock. No banks or business houses were robbed, nor were there any guards in attendance. The local hold-ups were of individual men after dark; the burglaries were petty-larceny affairs. Comstock fighters, in the main, were amateurs—local "tinhorn gamblers," ex-policemen, and saloon roughs. Bad whiskey and a handy gun caused most of the affrays. Roswell K. Colcord, ex-Governor of Nevada, wrote in the Mining and Scientific Press of March 11, 1922:

> Although Virginia City had a hard reputation, I know from personal experience that it was as orderly as a Quaker meeting in comparison with either Bodie in the late '70s or Aurora in the early '60s, for I lived during the roughest times in these places.

THE PLACERVILLE ROAD—THE GEIGER GRADE

Well-to-do people made frequent trips to San Francisco, to which they referred as "going down below" or "to the Bay." The Placerville road was the most popular, although the stages over Henness Pass carried fully one fourth of the passenger traffic. The ride over the Sierras was exhilarating, and the trip from Sacramento to San Francisco was a pleasant one as the river boats were very comfortable, almost luxurious. The stage ride over the Sierras was to Samuel Bowles, editor of the Springfield Republican, one of the events of his life. He said that words failed him in attempting to describe the scenery, and the perfection of the road and the skill of the drivers filled him with admiration:

EARLY-DAY TRANSPORTATION—The California Company's stage standing in front of the International Hotel in 1865. Those stages used the Henness Pass route to California, while the Pioneer Company's stages traveled the Placerville route. The sign on the brick wall reads: "California Stage Company, via Donner Lake, Forest City, Marysville, Sacramento." The sign marked '1' reads "Miners' Saloon."

I doubt if there ever was any staging at all comparable to this in perfection of discipline, in celerity and comfort, and in manipulation of the reins. * * * Think, too, of a stage road one hundred miles long, from Carson to Placerville, watered as city streets are watered, to lay dust for the traveler. With six horses, fresh and fast, we swept up the hill at a trot, and rolled down again at their sharpest gallop, turning abrupt corners without a pull-up, twisting among and by the loaded teams of freight toiling over into Nevada, and running along the edge of high precipices, all as deftly as the skater flies or the steam car runs; though for many a moment we held our fainting breath at what seemed great risks or daredevil performances.

His companion, Schuyler Colfax, Speaker of the Lower House of Congress, said in one of his speeches that it required more talent to drive a stage over the Sierras than to be a member of Congress.

A franchise to construct the Geiger Grade was granted by the Territorial Legislature on November 29, 1861, to Dr. D. M. Geiger and J. H. Tilton. Work was commenced early in 1862 and the road was in partial use by the end of that year. During 1863 it became a great highway. A winding highway of gentle grade, built in 1936, now runs down the mountain side from the old Five Mile House to Steamboat Valley and cuts out the steep portion of the old grade.

The Kings Cañon road, back of Carson, was completed in November 1863. The roads over Sonora Pass, Ebbett's Pass, Kit Carson Pass, and the Donner Pass were practically completed that year.

The Nevada Directory for 1863, on pages 6–8, prints tables of distances and stations over all of the Sierra roads, some of which were not then completed.

In the summer of 1865, when Bowles made his trip, the Pioneer stages were making the run to Placerville (101 miles) in twelve hours. Stages left Virginia City at 5 p. m. and passengers breakfasted at Sacramento. The stages from Virginia City to Sacramento were not making such fast time in 1863. The Pioneer line, via Placerville, advertised 22 hours to Folsom, 132 miles.

EARLY PROMINENT MEN

Nearly all of the men who became leaders on the Comstock arrived in 1863 or earlier. There is a distinguished list in the

Nevada Directory for 1862 and a longer one in the Directory for 1863.

The mine superintendents were among the first and remained the most constant. Chief among them were John W. Mackay, Philipp Deidesheimer, Robert Morrow, Charles Forman, Roswell K. Colcord, Samuel G. Curtis, C. C. Stevenson, Charles L. Strong, Adolph Sutro, H. G. Blasdel, Isaac L. Requa, Robert N. Graves, Albert Lackey, Harvey Beckwith, H. H. Day, Enoch Strother, and I. E. James—all remarkable men.

The leading lawyers in 1862 were William M. Stewart,[13] Moses Kirkpatrick, Richard Rising, J. Neeley Johnson, William H. Clagett, Jonas Seeley, Thomas E. Hayden, Dighton Corson, Martin White, and C. H. S. Williams. Eleven physicians were listed in the Directory, but nearly all of them soon departed.

The newspapers were handled by a talented group; The Territorial Enterprise by Joseph T. Goodman and Denis E. McCarthy, the owners from 1861; by William Wright ("Dan De Quille"), a kindly humorist, who served as reporter for thirty-two years, and by Mark Twain, a reporter for twenty-one months. Rollin M. Daggett, who arrived in the summer of 1862, years later became the editor of the Enterprise. Judge C. C. Goodwin, who became an editor of the Enterprise in 1874, was then living at Washoe City. The Virginia Daily Union, which proved a strong competitor to the Enterprise, was started on November 4, 1862, by John A. Church, James L. Laird, and Samuel S. Glessner—all able men.

Philip Lynch, one of the ablest and most active of the newspaper men, started the Gold Hill News on October 12, 1863. It lived for forty years. Lynch died in 1872 and was succeeded by Alf. Doten, another early newspaper man.[14]

The church was early represented in Virginia City by a number of notable men; Father Patrick Manogue, pastor of the Roman Catholic Church, later Bishop of Nevada, and, still later, of Sacramento; Rev. Franklin D. Rising, rector of St. Paul's

---

[13]William M. Stewart had served as Attorney-General of California; Neeley Johnson had been Governor; Daggett was one of the founders of *The Golden Era*, the leading literary magazine on the Coast for many years. Blasdel, Stevenson, and Colcord became Governors of Nevada.

The Nevada Directory for 1862 lists the names of 40 men who were prominently identified with the camp at a later period, and in the issue of 1863 lists nearly all of them and many more, but the depression of 1865 thinned the ranks considerably.

[14]The early newspapers prided themselves on printing only the news, and that briefly. Sensationalism was barred. A fatal fight between would-be bad men was told in a few inches; and the report of the trial filled even less space. On the other hand, there was no limit on a report of public school exercises, nor on the details of a Fourth of July procession.

Episcopal Church, both of whom arrived in 1862; and by Rev. Ozi W. Whitaker, Rising's assistant and successor, who did not arrive until 1863. Rev. Rising, a rare spirit, died a few years later. Rev. Whitaker became Bishop of the diocese of Nevada, and many years later Bishop of Pennsylvania and Presiding Bishop of the Church in the United States. The service these men rendered to the community and the State is memorable. They built churches all over Nevada, founded schools and other institutions, and were in the forefront of all good works. "A genial spirit of fraternization characterized the clergy."

Father Manogue was a giant of a man, with a rough-hewn face, a big heart, and a fine mind. He had been a placer miner in California before going to Paris to study for the priesthood and his life was marked by simplicity and straightforwardness. His parishioners numbered more than those of all of the other churches, and for twenty-four years he gave to them a measure of devotion such as only a great soul could offer and a great body endure. He was six feet, three inches in height and weighed two hundred and fifty pounds. "Soggarth Aroon" (beloved priest) they called him. In his cassock he looked as tall as a steeple. His strong influence was always on the side of law and order.

CHAPTER IV

Development of the Washoe Process—Early Mining and Improvements.

The Washoe process for reducing Comstock silver ores was a mechanical adaptation of the arrastra for fine grinding the pulp from the stamps, and of the Mexican patio process by amalgamating the silver and gold in steam-heated pans—accomplishing in six hours that which had required from four to six weeks by the patio process.[1] It was not devised at once, but developed from two years of experimenting by Comstock millmen, chiefly Almarin B. Paul, aided by the inventive genius of California mill builders. Paul did the experimenting and the inventors built the pans.

No sooner had the placer miners dug down to the solid quartz than they began to build arrastras in which to mill it. Fifteen were soon in operation at Virginia City, others at Gold Hill, and a few on the Carson River where they were run by water power. The arrastras built to work Gold Hill ores were fairly successful, owing to the high percentage of gold in the early ore. Those operating on the complicated ore from the Ophir bonanza lost fully one half of the values, which led to the adoption of the Freiberg process by the Ophir, the Mexican, the Central, and the Gould & Curry. Several patio yards were installed in connection with the Freiberg mills for reducing rich tailings and low grade ores, but the process was only partially successful owing to the lack of warm and continuous sunshine.

The California miners, except those working on a small scale and using arrastras, were crushing their ores with stamps and catching the gold on silvered copper plates or on blankets. It was found that some complex ores required a more sustained period of amalgamation, and, just prior to the discovery of the Comstock, two California inventors, Israel W. Knox and Henry Breevort,[2] working separately, conceived the idea of constructing

---

[1]The patio process was a combination of the arrastra for crushing the ore to a pulp and the patio for decomposing the silver sulphides and bringing about amalgamation. The moistened pulp, mixed with quicksilver and a little salt and sulphate of copper, was exposed to the heat of the sun on an open floor and intermittently turned with shovels (or placed in a circular enclosure and trampled by horses or mules) until the sulphides were reduced to chlorides and then to metallic state, which united with the quicksilver to form amalgam. The gold in the ore united readily with the quicksilver.

[2]Breevort later had a mill in Gold Cañon.

iron pans on the principle of the arrastra, but operated by machinery, in which to amalgamate the gold after the ore had been pulverized by the stamps. Breevort, although his plans were crude, proposed further fine grinding in the pans between iron mullers and bottom plates.

Paul built a 24-stamp mill in Gold Cañon, just below Devil's Gate, and Coover & Harris set up a little 8-stamp mill at Gold Hill.[3] Both installed "Howland's Portable Batteries," in which the stamps were set in a circle around a central shaft.[4] Coover & Harris had one unit of eight stamps, Paul three of eight stamps each. There was sharp competition to see which mill would be first, and both commenced operations on August 11, 1860.[5]

The original Knox pan was designed for amalgamation only, and the quicksilver and amalgam were drawn from the bottom, resulting in a heavy loss of suspended particles. Later, when grinding pans were introduced, the charge was passed into settling pans from which the quicksilver and amalgam were drawn. Later still, the settlers discharged into agitators, where a further small saving was made.

The next thought of Paul and of the mill builders was to improve the process. Paul wrote later: "In 1860 and 1861 none of us knew anything about milling silver ores. We all talked of the patio, the Freiberg, and the barrel.[6] The Freiberg mills of the Central and the Mexican were in operation in the spring of 1861; that of the Ophir in the spring of 1862; and during the latter year the great Freiberg mill of the Gould & Curry was completed.

In order to rival the patio, it was necessary to provide for fine grinding the pulp in the pans, and to raise the contents to a high

---

[3]Dr. E. B. Harris, of Virginia City, wrote an account of his enterprise for Thompson & West *History of Nevada*, p. 68 (1881). Almarin B. Paul became one of the most prominent and useful citizens on the Comstock.

[4]*U. S. Mineral Resources* for 1869, p. 666, illustrates "Howland's Rotary Battery" in a chapter on the stamp mills of the period. The Comstock soon adopted the plan of placing stamps in a row of batteries of five.

[5]The prices of the two mills, if correctly reported, were very dissimilar. Harris says he charged $100 a ton at first, then reduced it to $75. Lord (p. 85) says Paul contracted to mill 9,000 tons at $30 a ton "if the quartz would yield a surplus profit to the owners at this rate." Paul wrote three years later: "We could hardly work ore at an actual expense of $25 to $30 per ton, while now better work can be done for from $12 to $15 in milling expense—that is, in mills of 50 tons daily capacity." (*San Francisco Bulletin*, April 10, 1863.) The usual charge for milling during 1861 was $50 a ton; then reduced to $35 and to $20, and to $15 by 1865.

[6]*San Francisco Bulletin*, April 10, 1863.

degree of heat. Late in 1860, William H. Howland of San Francisco changed the form of the Breevort castings and constructed a strong serviceable pan 5 feet in diameter and 12 inches deep, with a heavy muller of cast iron made to revolve rapidly upon cast iron shoes, which provided the necessary fine grinding.[7]

A year later Howland devised a steam chamber to be attached to the bottom of the pan,[8] by which the pulp could be raised to any required temperature. Others led live steam into the pan, but the steam chamber proved the more serviceable. By 1862, progressive mill men had adopted fine grinding and steam heating, which Paul described as "the patio process perfected."[9]

The average charge of quicksilver was ten percent of the weight of the ore; a larger quantity being required for rich ores. The average loss of quicksilver was 1½ pounds for each ton milled. When rich ores were worked the loss of quicksilver and silver values was largely increased owing to the grinding of the sulphides into minute particles which floated off in the muddy water with the floured quicksilver. This stuff was called slimes, and was usually saved in reservoirs near the mills. While often very rich, the slimes were expensive to treat, and the values difficult to save.

The large lumps of ore were no longer broken with hammers (or by especially heavy stamps called "breakers") and fed to the stamps with shovels, but crushed in Blake rock breakers to egg size and fed automatically to the stamps, where it was crushed wet to 40 mesh or less. The automatic flow of the pulp was interrupted only between the stamps and the grinding pans. There the pulp flowed into wooden troughs to drain off the water; then shoveled upon a platform from which the pans were charged, again with shovels. The pulverized ore could not be allowed to flow from the stamps into the pans due to the excess water used in the stamps compared with the pans.

---

[7] The inventors soon devised various types of pans, which are described and illustrated on pp. 683–691 of *U. S. Mineral Resources* for 1869; also in Vol. 3, pp. 193–293 of *U. S. Exploration of the 40th Parallel* (1870) ; and in Lord's *Comstock Mining and Miners*, pp. 82–88, 117–121 (1883).

[8] Paul credits Howland with that invention. (*San Francisco Bulletin*, April 10, 1863.)

[9] Id.

Paul said in 1863, "I see much working, costing thousands of dollars, which could be demonstrated as a perfect humbug, a nonsensical expenditure of time and money. The more work is simplified, the more one comes down to solid common sense, with less of this scientific tomfoolery, the more money will be made." (*San Francisco Bulletin*, April 27, 1863.)

INTERIOR OF WASHOE PROCESS MILL

The stamps are at the right; the grinding steam-heated pans in the upper row; settling pans in second row; and agitators below to catch the values that escaped the settlers.

The Freiberg process, which involved dry crushing with stamps, chloridizing-roasting in ovens, and amalgamating in revolving barrels or in tubs, was a slow, intricate, and costly process. It was very successful in saving the silver (averaging 80 percent), but lost part of the gold. The Washoe process, on the other hand, saved the gold but lost part of the silver. The Freiberg was so expensive that the Ophir, the Mexican, and the Gould & Curry discontinued its use after their rich ores were exhausted. However, until 1870 it was the custom of the various mines to sort out their small amounts of high grade ore and send them to Dall's Freiberg custom mill at Franktown, in Washoe Valley,[10] for treatment at a cost of $50 a ton.

There were 75 mills in the Comstock region in 1863, a third of which could be spared. There were 19 in Virginia City and in Six and Seven Mile Cañons below; several of them using the Freiberg process. Gold Cañon was lined with 35 mills from Gold Hill to Dayton. There were 12 more on the Carson River from Empire to Dayton, and 9 in Washoe Valley and on Steamboat and Galena creeks below. The Gold Cañon and Carson River mills were reducing Gold Hill ores and employing the Washoe process. About a dozen arrastras continued to operate in the cañons and on the river.

EARLY MINING AND IMPROVEMENTS

That was a new and strange country. None of those early Comstockers knew anything about silver mines, and the ablest of them could only guess what the next day's work would disclose; besides, they knew little of quartz or lode mining. That kind of mining was still in a primitive stage in California, where the problems were comparatively simple. Those early Comstock miners, therefore, had no background; they had to learn by doing; faced by difficulties which steadily became more formidable than those encountered in any other mining camp in the world, before or since. If those men could have foreseen the interminable battles before them with bad air, floods of hot water, intense heat, and treacherous caving and swelling ground, their stout hearts would have failed. As it was, they grew with their tasks and became great miners from sheer necessity. The art

---

[10]Washoe City in 1863 was a virile, thriving town of 1,500—larger than Carson City. (*Directory* of 1863.) When the mills ceased to operate and the forests were cut down it sank slowly into oblivion.

of modern mining was acquired on the Comstock and then taught to the whole world.

The old theory that the Lode dipped westward into the range still prevailed in 1863, and substantial hoisting works remained along the croppings on the mountain-side above the streets.   In 1864, when it became known that the westerly inclination of the early ore bodies was a false one, and that the lode dipped easterly under Virginia City, new vertical shafts were commenced down on "D" Street and below.   The Gould & Curry and the Chollar, favored by steep side hills, extracted much of their ore through tunnels which also drained the mines.   They soon had to begin to sink shafts.

All of the early mines did their hoisting with large iron buckets, and when the Savage installed cages there was much argument over the utility of the two methods—the bucket having the advantage of hoisting water.   That was met by hanging a bucket below the cage until the increase of water in the second-line shafts compelled the introduction of Cornish pumps.   The Ophir, located on the drainage from Spanish ravine had a never-ending contest with water.

The Sides and the White & Murphy (which became the Con. Virginia) were drained in 1864 by the Latrobe Tunnel, 2,800 feet in length, which tapped the Lode on its dip at 700 feet.   A drift was run to drain the Central and the Ophir.   A similar tunnel to tap the Little Gold Hill mines was driven 800 feet and abandoned when their ore diminished and hard times came.

The mines with large ore bodies employed the square-set system, but the art of timbering shafts developed slowly—in fact, not until they began to sink second-line vertical shafts with three or four compartments.

The ventilation of the mines, so that men could live and work, was the most difficult of all problems from the beginning.   The air and the water began to increase in heat and foulness not far below the surface, and the men laboring at hand drilling, shoveling, and timbering where there was no circulation of air were quickly affected.   More men died from that cause than from any other during the early years.   The mine managers, who themselves had so much to learn, tried various expedients, but not until Root blowers were introduced in 1865 was there any improvement.   The various mines, with some exceptions, were not connected for years; meantime each had to solve its own problems.   The first real relief came when Burleigh mechanical drills, driven by compressed air, were introduced; the first being in the Yellow Jacket in 1872. The compressed air not only fed the drills of the miners, but ventilated workings and ran blowers in distant places and small underground hoists.

A. S. Hallidie, that inventive genius of San Francisco, devised the flat woven-wire rope in 1864, which became the standard on the Comstock. The width was about four inches and the thickness from one half to five eighths. Prior to that the mines used hempen ropes. Safety clutches were soon installed on cages to stop them automatically in case the rope broke.

Black powder was the only explosive until giant powder (dynamite) began to come into use in 1868.

CHAPTER V

The Depression of 1864—Arrival of William Sharon—Civil War Spirit—Nevada Becomes a State—Stewart Elected United States Senator.

The end of the year 1863 arrived with no marked change in affairs. Some of the smaller mines had closed down, there was less general employment, money was becoming scarcer, and many had departed for California—among them a substantial number of the baser element.

Stocks in the leading mines had held up well: Gould & Curry hung around $5,000 a foot, Ophir $1,650, Savage $2,650, Yellow Jacket $950. The stocks of some of the other mines along the Lode had suffered a greater decline, but nothing alarming. The prices of wildcats had suffered most.

Nevertheless the market was shaky. The dividends from the rich mines had fallen far below expectations, while the expenditures had been enormous. There had been no return whatever from all of the others upon which millions had been expended. It was rumored that the Gould & Curry and the Ophir bonanzas were playing out on the lower levels. The market needed but a push to bring it down and that was furnished by reckless, bloody, extravagant Aurora, early in 1864, when its stocks crashed following the failures of its rich surface ore bodies on the Del Monte and the Wide West at the depth of 100 feet.[1] Aurora was finished as a great mining camp and declined rapidly.

The depression on the Comstock deepened throughout 1864. "We are in the midst of unprecedented and unexpected hard times," wrote Charles De Long, a prominent attorney, in May 1864.[2] The Virginia Daily Union of May 13 printed a long editorial complaining of dull times:

> Merchant and miner, banker and broker, lawyer and doctor, join in the chorus and echo the oft-made assertion that never since 1860 has Virginia been so dull as during the last few weeks. * * * Stocks are down; money is high, and almost impossible to obtain even at the high-

---

[1] The crash in Del Monte stock in 1864 caused many failures among the brokers of San Francisco. King's *History of San Francisco Stock Exchange*, p. 21 (1910).

[2] *Virginia Daily Union*, May 25, 1864.

est rates. * * * From Austin and Aurora and Humboldt the same cry of dull times comes up unceasingly.

The complaint was general, and every reason given for it except the real ones—that the boom was a California brainstorm and that known ore bodies were failing. Yet men continued to look for new discoveries and a better market until the panic of 1865 blasted their hopes.

ENTER WILLIAM SHARON

The year 1864 was long remembered as the date of the arrival of William Sharon, who, with the powerful support of William C. Ralston and D. O. Mills in San Francisco, was to be the czar of the Comstock for seven years, from 1867 to 1874.

Ralston, the most generous, popular, and daringly enterprising banker in San Francisco for many years, was interested in the Comstock mines from the beginning. He was made treasurer of the Ophir and the Gould & Curry when they were incorporated in 1860, and thereafter became treasurer of nearly all of the leading mines. He had been a partner for several years in the banking firm of Donohue, Ralston & Co., from which he withdrew early in 1864. In June of that year he organized The Bank of California, with a capital of $2,000,000,[3] and a strong representative group of incorporators and stockholders. D. O. Mills became first president. Ralston, as cashier and manager, and later as president, dominated its affairs until his death in 1875.

The new bank opened its doors for business on July 5, 1864, and Ralston, with four years connection with the Comstock and a close acquaintance with local conditions, opened a branch bank in Virginia City with William Sharon as manager.

Sharon, a cold, vain, grey-eyed little man, topped-off with a tall plug hat, was forty-three years old when he came to the Comstock. Cynical and worldly-wise, he was the very antithesis of the men with whom he was to deal. He had been admitted to the bar, had been a merchant and trader, and, in San Francisco, a politician, real estate speculator, and, finally, a mining stock speculator, in which he lost his fortune. He was ripe for a new adventure.

D. O. Mills was a banker and businessman of Sacramento;

---

[3]This was in gold—equivalent to double the amount in greenbacks. The Pacific Coast did business on a gold and silver basis all during the greenback period, from 1862 to 1879.

reserved, cautiously venturesome, of rare financial ability and excellent character. He removed to New York City in 1876, lived long, and became one of the wealthiest men in the United States.

Alvinza Hayward, who joined Sharon and Mills in the Comstock venture in 1867, had already made a fortune through the ownership of the famous Eureka gold mine at Sutter Creek, California. He was a fine-looking, agreeable man, who remained faithful to his associates and shared in their good fortune until he and Superintendent J. P. Jones conspired to take control of the Crown Point away from them soon after its bonanza was discovered.

Sharon began to lend money on mills and other property at 2 percent a month, in competition with the ruling rate of 5 percent, and found eager borrowers. While no one anticipated that the depression would deepen until it culminated in the panic of December 1865, Sharon took the risks of an adventurer in lending the bank's money so freely when the bottoms of all of the early ore bodies had been found and no one knew whether others would occur below. The market had been falling steadily throughout 1864, and by the time the agency was opened for business, early in November, the stocks of the leading mines had lost two thirds of their value of the year before. Sharon might well have paused, even though his loans did not imperil the bank.

Both Ralston and Sharon were daring men and doubtless planned more than banking, but it is inconceivable that either looked forward to the almost complete control of the Comstock which followed. The hard times of 1865 and the halting recovery thereafter created the opportunity, and, indeed, forced it upon them as mills and other properties fell into the bank's hands.[4] Fortunately the Comstock revived in 1866 and enjoyed three years of moderate prosperity, during which Sharon made himself master of the Lode. The stock market was not very active during that period, there were no longer any strong competing interests, and the bank's money paved the way. By May 1867 the bank had seven mills on its hands, which Sharon and his associates could use, as they already had control of the Yellow Jacket and the Chollar-Potosi and were planning the control of other large producers. Accordingly, in June of that year, Sharon, Ralston, Mills, and Hayward took over the mills and formed the Union Mill and Mining Company. Two years later they

---

[4] Lord's *Comstock Mining and Miners*, p. 246 (1883).

controlled all of the leading mines and had seventeen mills.[5] If Sharon wanted a competitive custom mill he starved it into submission by withholding ore. His system of private milling by those in control of mines became a Comstock custom.

The Bank of California increased its capital to $5,000,000 (gold) in 1866, and became and remained for many years the foremost financial institution on the Pacific Coast.

The control of the mines not only insured a monopoly of milling but enabled Sharon to dictate highly profitable milling contracts to the various boards of trustees (directors) elected by him and his friends. The management of the mines also was in his hands and he controlled them as if he were sole owner.

Fortune again favored the combination, which became known as the "Bank Crowd," in that even more depressing year of 1870, when the last of the early ore bodies was nearing exhaustion and ruin threatened them and the bank. They had loaned three fifths of the bank's capital on the Comstock and attendant industries,[6] in obtaining complete personal control—and the lucky discovery of the Crown Point bonanza toward the end of the year was all that saved them.

CIVIL WAR SPIRIT—NEVADA BECOMES A STATE

Full telegraphic news of the progress of the Civil War was a feature in all of the papers. War spirit was at fever heat; volunteers were enrolled daily. The Unionists were largely in the majority, but the "sesesh," or copperheads as they were called, made up in violence of speech what they lacked in numbers, although the leaders were careful not to become so offensive as to invite a sojourn at Fort Churchill in the Carson Desert, where the "guests" were invited to pack bags of sand around the parade grounds for exercise.

The Territory seethed with political excitement during the fall of 1863, over the election of delegates to a convention that was to frame a constitution for the proposed State. The contest was between the Union Party and the Democrats and the former was overwhelmingly in the majority. That constitution was rejected by popular vote, because of the provision taxing the mines and because the list of state officers to be voted upon at the same time

---

[5] Lord's *Comstock Mining and Miners*, pp. 246–248 (1883).

[6] So Sharon told Lord. *Comstock Mining and Miners*, p. 279 (1883).

The only virtue Sharon claimed was that he told the truth. It was also said that he was loyal to his friends; women were his bane.

was opposed by a large number of patriots who aspired to the offices. The people then elected delegates to a new convention which met in July 1864. This also was overwhelmingly Union; but one Democrat was elected. A homographic chart[7] shows all but three to be American born and from northern States, and that all but four came from California—the majority of them having arrived in '49 or soon thereafter. The average age was thirty-five. The constitution was ratified, and the State admitted October 31, 1864. Of the eighty-four State officers elected, including members of the Legislature, seventy-seven were native Americans.

STEWART ELECTED UNITED STATES SENATOR

The year 1865 opened with a new State of Nevada fully organized, including the selection of two United States Senators. The dominating personality of William M. Stewart overshadowed all other candidates. He was not only the foremost lawyer, but an all-round dictator, including the leadership of the Republican party. When the first State Legislature met on December 15, 1864, he was promptly elected. Former Governor James W. Nye was chosen as his associate.[8] Thereafter, during bonanza days, those honors were sought by men of great wealth and became bargain-counter affairs.

Stewart was as able and diligent a United States Senator as he had been as an attorney and was reelected in 1869. One of his early accomplishments was to secure the passage of the first Federal Mining Act, known as "The Law of 1866." He was also the father of the later and more comprehensive mining law known as "The Law of 1872." He ceased to be the ruler of the Comstock after his first election to the Senate. His mining and milling enterprises had not been profitable, there was no longer any mining litigation of importance, and his interests lay in Washington. When his second term expired in 1874 he did not contest with

---

[7]Printed on pp. 81, 86 of Thompson & West *History of Nevada* (1881). The total vote cast for and against the Constitution was 13,655.

When H. G. Blasdel, an able though somewhat uneducated mining man, was a candidate for governor he received the following endorsement from William "Billy" Woodburn, a prominent lawyer, in a public speech: "Ladies and gentlemen," he urged in his high-pitched voice, "can it be possible that the people of the sovereign State of Nevada propose to elect to the highest office in the gift of the State a man who spells God with a small 'g,' Christ with a 'K' and Blasdel in capital letters?" Blasdel proved an excellent executive and was reelected four years later.

[8]Thompson & West *History of Nevada*, pp. 88, 89 (1881), tells of Nye's belated selection.

Sharon for reelection, but turned to mining—in Bodie, in the Panamint Range, and elsewhere—always without a success. When he was promoting the Noonday mine at Bodie in 1879 and 1880 and building a costly mill, this writer, then the messenger boy in the local telegraph office, often delivered messages to him.[9] His hair and luxuriant beard were white as snow and he walked like a cathedral in motion, always alone. He was only fifty-five. About the year 1884 he formed a law partnership in San Francisco with William F. Herrin, counsel for the Southern Pacific Railroad. Although Stewart had not lived in Nevada for many years, he returned in 1886 and was again elected to the United States Senate (succeeding James G. Fair), and for two additional terms, with the assistance of the Southern Pacific Railroad. Nevada was ably represented in the United States Senate during those years, where Jones and Stewart served as the spearhead in the contest for the remonetization of silver.

---

[9] *The California Historical Society Quarterly* of March 1925 prints an article by this writer on "Bodie, the Last of the Old-Time Mining Camps."

"C" STREET, VIRGINIA CITY IN 1865, LOOKING NORTH

Note the teams of oxen hauling freight wagons—part of the animals lying down, as is their habit when resting. It is probable that they were hauling wood or lumber from Washoe Valley.

The building on the left, with the porches, was known as McLaughlin & Root's. Beyond that stands the Gillig & Mott building, with the International Hotel (with the high ridge roof) still further along the west side of the street.

The first tall structure on the right is the Enterprise Building, erected in 1863. Beyond that the Medan Building. South of the point from which the photograph was taken "C" Street was lined with substantial buildings for two blocks.

All of the buildings shown in this print were destroyed in the great fire of October 26, 1875. Those that replaced them were not so impressive, with the exception of the International Hotel.

Cedar Hill and Lone Rock in the distance.

CHAPTER VI

1865, An Emotional Year — Lee Surrenders and Everybody Drunk—Lincoln's Death and Funeral Procession.

The depression of 1864 was followed by a dull and trying winter; yet, such is the hopefulness of mining camp men, that the Comstock looked for new discoveries and better times in 1865. Little did they know that it was to be the worst, and that half of the population would be gone before the year ended.

Nothing better illustrates the youthfulness of spirit of those Comstockers than the way in which they received the news of Lee's surrender, and, on the heels of that, the tidings of Lincoln's death.[1] The Daily Union of April 12, 1865, tells of the first:

Day before yesterday morning all was quiet—everything was going along as usual here in Virginia, and everybody were at their customary avocations. Suddenly, about 11 o'clock, spread like an electric shock, the glorious news just flashed across the continent, of the surrender of Gen. Lee, together with his entire army, to our victorious Grant. All of the papers had out extras containing the great news. At 12 o'clock one bell after another commenced ringing, until at length about all the bells of the city rang, including every one of the church bells. Numerous steam whistles also chimed in to swell the din which was kept up for about two hours steady. The scene in the streets defies description. It seemed as though everybody had suddenly gone mad. People were rushing hither and thither, flags were being hoisted, and all loyal men seemed to feel that now was the time when they should show their colors. The Enterprise and the Union offices both had their flags out, and from every flagstaff, and strung across the streets everywhere, floated the good old Banner of the Stars. We even noticed several of the Copperhead order who entered with zeal into the general manifestations of joy. There were many, however, who looked scowlingly and sullenly on, inwardly cursing the news, and all who rejoiced thereat. But what did Union men care for

---

[1]Lee surrendered on April 9, 1865; President Lincoln was shot by John Wilkes Booth on the 14th and died the next day.

them or their feelings? * * * The military companies
at once donned their uniforms and paraded the streets.

EVERYBODY WILD DRUNK

The saloons did a tremendous business. They were
crowded full from the commencement, and everybody
drank cocktail after cocktail in the very joy of their
hearts; they drank to one hero after another; they
drank to "Old Abe"; they drank to the "Old Flag";
they drank anyhow. No such drinking was ever before
seen anywhere. In less than three hours the majority
of the men in the city were crazy drunk. Friend meeting friend, drinks came in as a matter of course, and
this operation was repeated every five minutes, or oftener. As an instance of the amount of liquor drank, in
one of the principal saloons there were over fifteen hundred drinks sold. Men were drunk that never were
under the influence of liquor before, and scores were to
be found laid out singly and in heaps almost anywhere.
Business was entirely suspended, and the printers, editors, reporters, and proprietors being all drunk, no
papers were issued. No newspaper in Virginia City
was issued for two days.

HOW LINCOLN'S DEATH WAS RECEIVED IN VIRGINIA CITY

Five days after the celebration over Lee's surrender the Comstock was stricken by the news that President Lincoln had been
assassinated. Every column of the next issue of the Virginia
City Enterprise[2] (eight pages of news and advertisements) was
black-leaded as a badge of mourning. An editorial told how the
news was received. Only on the Comstock could the tragedy
have so completely suspended every activity:

When the news of the death of Lincoln reached Virginia City on Saturday morning, April 15, almost the
entire population gathered in the streets and in low
tones talked over the calamity which had just overtaken
the Nation. There was no loud talking, drunkenness or
bluster, but every word was expressive of mingled grief
and wrath—deep and intense.

The church bells tolled all day.

[2]The *Daily Territorial Enterprise*, April 18, 1865. This is one of the few sur-
viving copies of the newspaper for 1865.

The city was draped in black. Every store was closed, the banks all closed, the courts at once adjourned, and the doors of every saloon and whiskey shop were locked and draped in mourning.

The quartz mills all shut down and the hoisting works stopped.

The two theatres, Maguire's Opera House and the Music Hall omitted their usual Saturday afternoon and evening performances.

The quiet through our city was such as has never before been seen.

In very few instances men expressed joy at the news. They were roughly handled and lodged in jail.

Elaborate arrangements were made to hold a public funeral on Thursday, the 19th, in which every organization in town and all of the leading men took part.[3]

[3] The *Virginia Daily Union* of April 20 printed a detailed account of the funeral procession and of the later ceremonies at Maguire's Opera House.

CHAPTER VII

The Panic of 1865—Comstock Production and Profits from 1859 to 1866—Stock Devilment.

In 1865, six years after its discovery, the Comstock appeared to have seen its best days. The ore bodies along the Lode had terminated at an average depth of 500 feet and no new ore of importance had been found. Although the yield of the mines had increased from $12,400,000 in 1863 to $16,000,000 in 1864, and returned the same amount in 1865, men knew only too well that the remaining ore in sight would not last over two years. They knew also that although the mines had made a total production of $50,000,000 during the first six years, the Comstock mines as a whole had failed to show a profit—in fact entailed a loss; the assessments and expenditures exceeded the dividends. High costs of operation, wasteful and extravagant management, and "millions uselessly and foolishly expended in exploration work," together with vexatious and costly litigation, had consumed that splendid output.

The general average of the 1,074,078 tons produced by all of the mines from 1859 to 1866 was $45.48 a ton. Milling costs, which were necessarily high during the early years, had absorbed nearly one half of the product, one fourth or more had gone for general expense, and a little less than one fourth was paid in dividends.

The operations of the leading producing mines had been disheartening. The rich ore in the Ophir and the Gould & Curry was only a memory, and the remaining low-grade ore would last but a year or two. The rich upper ore body of the Savage seemed to be exhausted. It had produced $3,600,000 from its extension of the Gould & Curry bonanza, had paid only $800,000 in dividends, and was in debt nearly $500,000. The Hale & Norcross had spent $350,000 without finding ore. The Chollar and the Potosi, after four years of disgraceful wrangling in the courts, costing $1,300,000, had compromised by uniting as one mine. The case was referred to as Jarndyce v. Jarndyce.

The little mines at Gold Hill (including the Imperial and the Empire) were on the decline, although continuing to produce a large volume of ore of moderate grade at a profit. The fine ore body discovered by the Yellow Jacket in 1863 at the north end of the mine at 180 feet appeared to have given out at the depth of 360 feet. It had produced $3,880,000 in three years, paid $420,000 in dividends, levied $330,000 in assessments, and was in debt over $300,000. The rich near-surface ore body of the Belcher, which stood in the west ledge, had suddenly given out at the depth of 300 feet. The mine had the best record of any of the producers, with the exception of those located on the Gold Hill bonanza. It had paid $421,200 in dividends out of a production of $1,400,000. Both the Overman and the Caledonia had produced considerable ore from small scattered ore bodies, at no profit.

Mining men and speculators were dismayed and dumbfounded to see all of those early bonanzas, from the Ophir to the Belcher, play-out at the depth of 500 feet or less. All of their hopes and talk had been of the inexhaustibility of silver mines—of which South America, Mexico, and Central Europe provided such shining examples. Nor could anyone have foretold the catastrophe. Many of the prodigal expenditures of the early '60s can be excused because of the prevailing belief that the ore bodies would last indefinitely.

Fifty other mines along the Lode, or in its immediate vicinity, whose stocks had been freely bought on the exchange at high prices, had nearly all closed down, after levying millions in assessments. Four hundred wildcat companies, whose stocks had been bought and sold on the local exchanges at extravagant figures during 1863, forfeited their charters and were soon forgotten.

Many thought that the Comstock was finished as a great mining camp. Baron Von Richthofen's timely report heartened the miners somewhat with the assurance that the Comstock was a true fissure vein which would extend to great depths and bear other large ore bodies, although he doubted whether they would be as rich as the near-surface deposits.

The market value of all of the Comstocks had fallen from $40,000,000 in 1863 to $12,000,000 in the summer of 1865, which men thought was the bottom. But the worst was yet to come. In October a panic seized the speculators, and on December 15 the total market value of all of the mines on the Lode had fallen to $4,000,000—notwithstanding the fact that they had produced $16,000,000 during that year. The decline in the value of leading stocks was appalling. Gould & Curry had fallen from $6,300 a foot in July 1863 to $1,650 a year later, and to $800 on December 15, 1865; Ophir from $2,100 to $250; Savage from $3,500 to $700; Yellow Jacket from $1,200 to $250; Chollar and Potosi from $1,000 to $117; Belcher from $1,500 to $130. These were the leading producing mines; the stocks of the nonproducers along the Lode had depreciated still more.

The average depth of the main shafts along the Lode at the end of 1865 was only 450 feet. That of the Bullion was 600 feet, and of the Gould & Curry nearly 600. These shallow depths were due to the fact that the first shafts were sunk along the croppings, and that the second-line vertical shafts were not commenced until some time in 1864.

Ten thousand people left the Comstock during 1864 and 1865; many returned to California; others sought fortune in the placer mines of Idaho and Montana; not less than 3,000 went to the new camp of Summit City, located eight miles north of Cisco on the summit of the Sierras. Machinery, mills, houses, and furniture were moved away to other camps—especially to Summit City, "the new Comstock," which was to be a flat failure.[1]

The whole Territory was involved in this crash of 1865. The value of hundreds of stocks and thousands of would-be mines melted into thin air.

In justice of those old-timers, it should be said that the cost of mining on the Comstock throughout its history was enormous. Supplies of all kinds except lumber were brought from California at great expense. Although those barren hills provided only a meager and in part mineral-polluted supply of potable water, the Lode was saturated and nearly all of the mines battled with water from the beginning. The Lode was so unstable that all openings required heavy timbering, which was constantly renewed, so great was the crushing power of the swelling porphyry and clay. Lumber was then $60 a thousand. Labor was the highest paid in the world. Milling costs, up to 1865, were necessarily high, whether ore was reduced in company mills or by custom mills. Nor had the latter been profitable, with few exceptions. In 1865 milling costs had become so reduced and extractions so much

---

[1]Summit City had an extraordinary career. Five or six thousand people rushed there in 1865 and 1866 and built a substantial city of homes, business houses, saloons, hotels, etc. It even had a stock exchange. Hundreds of mines were located and seven small mills erected. Some small deposits of rich gold ore were found upon and near the surface, but the prominent veins contained little save low-grade, rebellious ore. Suddenly in the fall of 1866, nearly the entire population departed, leaving practically all of their belongings; and well they did, for the city was buried under twenty-five feet of snow that winter, leaving the remaining inhabitants to communicate with one another by means of tunnels dug under the snow. For years that town stood, deserted and decaying, as if visited with a plague. Now the site is almost obliterated.

improved that custom mills were paying $1 a ton for old dumps which would average $15.

Millions of dollars had been spent in sinking shafts and driving tunnels on the theory that the Lode dipped westward into the mountains. Wasteful litigation had consumed an amount equal to one fifth of the production of the Lode up to 1866.

The dividends paid from 1859 to 1866 amounted to $11,375,900, and the assessments, as usually reported, totaled $5,215,118. But the latter tells only part of the story. There were 400 other companies, mostly wildcats, operating in the vicinity of the main Lode during those years, whose stocks were called on the local exchanges and in part on the San Francisco Exchanges. Their assessments, ranging from $1,000 to $500,000 each, have been tabulated at $5,000,000. In addition, not less than $1,000,000 was expended on mines by individuals, and another $1,000,000 was paid for mines by the original purchasers and incorporators. Including all assessments and expenditures, the mines in the district showed a loss of not less than $1,000,000 during the first six years of operation, and it may have amounted to $2,000,000. A table giving the production of the various mines from 1859 to 1882, together with tonnage, value, assessments, and dividends, will be found in the Appendix.

From an economic standpoint the Comstock Lode was butchered from the first by the location of many small claims, and by their retention as independent properties by the purchasers from California who came in 1859 and 1860. The craze for stocks and mining corporations that followed soon placed the situation past remedy.

The evils that beset the Comstock Lode are vigorously set forth by Rossiter W. Raymond in U. S. Mineral Resources for 1868, page 51:

> Nearly $100,000,000 have been extracted from that one Lode within the past nine years, yet the aggregate cost to owners had been almost as much. The reason is simple. Unnecessary labor has been employed, and vast sums of money wasted in extravagant speculations and litigation; *and the root of the whole evil lies in the system of scattered, jealous, individual activity, which has destroyed, by dividing the resources of the most magnificent ore deposit in the world.* Thirty-five or forty companies, each owning from 10 to 1,400 feet along the vein, and each (almost without exception) working its

own ground independently; 40 superintendents, 40 presidents, 40 secretaries, 40 boards of directors, all to be supplied with salaries, or, worse yet, with perquisites, or, worst of all, with opportunities to speculate; an army of lawyers and witnesses, peripatetic experts, competing assayers, thousands of miners, uniting to keep up the rate of wages; these things explain the heavy expense of Comstock mining.[2]

Owing to those unfortunate conditions, and to the vast amount of development work done below the 1,000-foot levels, only five companies, the Con. Virginia, the California, the Kentuck, the Crown Point, and the Belcher eventually paid more in dividends than they collected in assessments.   Not less than $50,000,000 was spent in deep mining from 1872 to 1886, when the last pumping shaft, the Combination, closed down.

STOCK DEVILMENT

The Comstock mines early became a synonym for stock devilment, for which the brokers and manipulators in San Francisco have been held solely responsible.   True, theirs is the chief responsibility, but the gambling public must bear its share of the blame.   The system early developed from a speculative investment in mining stocks into a gamble—then into a lottery.

Lord, referring to this and later periods, states that certificates of stock became tickets, and the holders won or lost on the turn of the wheel.   "A few prominent capitalists," he says, "purchased the control of mines for the sake of dividends, and the profits of milling ore, but the great body of holders bought their shares to sell at an advanced price."[3]   "A well-managed 'stock deal' was as acceptable to most holders as an actual development of ore."[4]   "So far as the Comstock mines have furnished opportunities for stock deals, their discovery and development have been a curse to the Pacific Coast, which all candid observers have recognized.   The Lode which was a boon to the thousands who found in it opportunity for persistent and useful work, was also a bane to the thousands who converted it into an instrument for trickery and passionate gaming."[5]

---

[2]Rossiter W. Raymond was a "famous personality in American mining affairs for half a century, and exerted a wide influence by reason of his intellectual vigor and high character." Rickard's *History of American Mining*, p. 123 (1932).

[3]Lord's *Comstock Mining and Miners*, p. 318 (1883).

[4]Id., p. 286.

[5]Id., p. 319.

Lord might have added that the control was often bought and utilized for the purpose of "rigging the market," and for the salaries, luxurious offices in San Francisco, and other perquisites that opportunity afforded. It should be said, however, that if the San Franciscans milked the Comstock, they also fed it. Not less than four fifths of the assessments collected by the mines throughout their history was contributed by stockholders in California and elsewhere. To the people on the Comstock, where money was expended, an "assessment mine" was almost as helpful as one that paid dividends; but they were the victims as well as the beneficiaries of the widespread mania, for they became inveterate stock gamblers.

In criticizing those who participated in that mining stock gambling—both dealers and players—we should remember that men are to be judged by the standards of the times in which they live. Nearly everybody gambled in mining stocks in those days—and knew they were gambling. The aim of all was "to beat the game." The men who won, whether dealer or players, were admired and envied. It was a mark of distinction to be called a "big manipulator." Those of us who lived through that hectic Comstock period, and, as well, through more recent mad speculative eras which have left millions in this country ruined and hopeless, are unable to see any difference between the two periods, either in morals or in methods. It may be well to recall that the methods of the San Francisco stock manipulators were but a pale reflection of the brigandage of the New York manipulators, headed by Daniel Drew, Jay Gould, and Jim Fisk, during the period immediately following the Civil War. Don Seitz called it "the dreadful decade." "It was a time of organized lawlessness under the form of law; of reckless gambling, with corporate securities as tools; of panics and of 'corners' in stocks and gold."

Editor Fred MacCrellish of the *Alta California*, the leading journal of San Francisco in the '50s and '60s, sought to lead a crusade against stock gambling in 1870 and 1871. In one issue, he said: "This stock jobbing business is simply gambling, and of the most demoralizing kind; for, unlike card playing it is pursued openly and has been regarded as respectable, just as lotteries once were. It is worse than card gambling, because the players are not upon an equal footing; its demoralizing influence more widely permeates all classes of society without regard to sex or age, and it breeds an increasing crop of professional liars whose business it is to entrap honest but credulous people." He endeavored to enlist the aid of the press and the pulpit and was making some headway when the discovery of the Crown Point bonanza created another wild market. (*Alta California*, July 16, 1871.)

CHAPTER VIII

The Law Governing Mines—Litigation.

American mining law is peculiar in permitting a miner to follow his vein on its dip into the earth even though it may penetrate beneath other men's mining claims. Other countries allow the miner a larger claim but with vertical boundaries beyond which a vein cannot be followed unless the miner has acquired the adjoining ground. The American system had its origin in the early placer camps of California. There was no Federal law in those years permitting miners to go upon public lands and locate mining claims. The miners were in fact trespassers, and were threatened from time to time by Washington, but their numbers became so great that all thought of interference was abandoned.

The miners solved that problem in true American fashion by adopting rules in various mining districts prescribing the size of placer claims and the conditions under which they could be held and worked—the size of the claims depending at times upon the richness of the deposit and the number of miners clamoring to share in it. But placer workings often led to rich little quartz veins, which presented a new problem. District rules had allowed a placer claim so many feet square, but veins dipped into the earth at varying angles and some of them would soon pass beyond a vertical boundary. New rules were then made allowing a miner to locate a vein for about 200 feet in length, and permitting him to follow it into the earth "with all of its dips, spurs and angles," which became known as "the extralateral right." This idea may have related back to the old practice in the lead mines of the Forest of Dean, England, where the right to follow veins was permitted, or perhaps to the Hartz in Germany.

That solution seemed a simple one at the time, but after the system was incorporated into the Federal mining laws of 1866 and 1872, and after the great development of the quartz mining industry—too late to make a change—it appeared that those early miners had opened a Pandora's box. Veins as a rule are most irregular in strike and dip, and the technicalities and complexities that arose in the administration of the law almost pass belief, as appear in Lindley's American Mining Law (3d Ed.) in six large volumes.

The Comstock mines were located under district mining rules

similar to those of California.   A locator could claim 300 feet along a vein (later reduced to 200 feet), and a group could locate a single claim with the same allowance for each member, an additional claim being allowed for discovery.   The width of a claim was not prescribed.   The locator was entitled to the full width of the vein, which on the Comstock turned out to be 1,000 feet in places.   The end lines, which define the length of the claim along the vein, were the controlling factors; within those lines, which are theoretically extended downward, a miner could follow his vein on its dip indefinitely.   The end lines also formed definite boundaries between claims beyond which a miner could not pass. When an ore body pitched through an end line into an adjoining claim (as when the Gould & Curry bonanza passed into the Savage, and the Crown Point bonanza into the Belcher) that portion of it became the property of the mine into which it extended. The Little Gold Hill mines, which were located over the same large ore bodies, worked in harmony by ceasing to extract ore when a neighbor's line was reached.

During the first few years the Comstock Lode was thought to dip westerly under the Virginia range, because the Little Gold Hill mines were extracting their ore from the Old Red Ledge, which dipped westerly at 45 degrees; while, in the Virginia City section, the Ophir, Gould & Curry, and the Potosi ore bodies which lay upon or near the faulted east wall, inclined to the west for the first few hundred feet.   Melville Atwood, an experienced miner from Grass Valley, warned the Ophir Company in 1859 that the footwall dipped east and that the Lode would follow.[1] If the Ophir had then bought up the footwall claims it would have saved a million dollars and years of litigation, chiefly that notorious conflict with the Burning Moscow.

It took the miners several years to work out the structure of the wide upper portion of the Lode, which was far from being clearly defined on the surface, as it was largely worn away and covered with debris from the mountain sides above.   The occasional large croppings, which stood along the west or footwall side, consisted of low grade and more resistant quartz.

The Comstock was not then thought of as one great vein or

---

[1]Atwood wrote: "The course of the Ophir vein appears to be a few degrees west of magnetic north, dipping westward at an angle of about forty-eight degrees, but the dip being a false one, you may not sink far before it will change." He said that the large outcrop on the west, which has an easterly dip, is the true footwall of the Lode, and the Ophir vein will follow it when they come together. (*Sacramento Union*, November 10, 1859.)

Lode, but as a wide mineralized zone in which there were a number of parallel quartz veins separated by belts of porphyry and sheets of clay. Under the early mining rules a miner was entitled to locate on but one vein; other adjoining parallel veins were open to location by others. The several veins which appeared on the surface of the Comstock were located as separate mining claims more or less conflicting with one another, which brought about much of the litigation that bedeviled the camp for nearly five years. The deeper workings in the various mines then showed that these parallel veins united at the depth of 500 feet or more to form the normal Comstock Lode which dipped eastward at an average of forty-five degrees, and the courts awarded the full width of the Lode on the surface, and underground as well, to the first locations made, which were those owned by the large companies. This controversy was locally known as the "many ledge theory against the one ledge theory." Local feeling as expressed in the newspapers favored the many ledge owners as they were far more numerous. The question was settled in favor of one ledge by the report of Referee John Nugent in August 1864, after its adoption by the courts. That decision settled also the right of the owners of the main Lode to follow the Lode on its dip, which had been claimed by each of the separate parallel veins.

It was largely upon the basis of the regularity of the strike and dip of the Comstock Lode that Senator Stewart was enabled— with the support of other western Senators and Congressmen—to pass the congressional mining laws of 1866 and 1872, which perpetuated the extralateral right and provided for the patenting of mining claims. Until 1866 the miners were without any legal rights upon the public domain throughout the West.

AN ORGY OF LITIGATION

The Comstock was a happy hunting ground for lawyers during the early years. They flocked there like buzzards after carrion, and engaged in an orgy of litigation over mining claims, much of it incubated in blackmail and reeking with perjury. The crude and indefinite early notices of location, and the practice of jumping claims, brought on many lawsuits. The question whether the Lode was one great vein or a series of parallel and independent ledges aroused the bitterest litigation.

All of the leading mines were involved in fifteen or more lawsuits, and others in nearly as many. Nine companies had 359 cases on their hands, nearly half of which they brought themselves against adjoining claims. The court calendars were clogged with cases. "Our principal industry is litigation," wrote an engineer of that period. Some of the underpaid "carpetbagger" Territorial judges were compliant and all were forced to resign the following year, largely through the efforts of Stewart after he lost the Chollar case. Statehood, in 1864, enabled the people to elect their own judges,[2] who, aided by the depression, speedily cleared their dockets of a multitude of pending cases, few of which came to trial.

William M. Stewart, afterward United States Senator, was attorney for the principal mines in the early years and dominated the Comstock. "He towers among men like the Colossus of Rhodes," wrote an admirer, which brought the retort from the Gold Hill News, "and has as much brass in his composition." He was a big, highhanded, domineering man, accustomed to driving his cases through the courts, but met his match in his first trial against A. W. "Sandy" Baldwin, an able lawyer who arrived in the fall of 1862. After the court had sustained several of Baldwin's objections to Stewart's methods, Stewart turned savagely: "You little shrimp, if you interrupt me again I'll eat you." To which Baldwin quietly retorted, "If you do you'll have more brains in your belly than you ever had in your head." Stewart soon took him into partnership.

Rollin M. Daggett, in telling of the notorious Yellow Jacket case, which was tried in June 1863, illustrates his gift for satire in the following paragraph, which is far too sweeping:

> The bar was noted alike for its solid attainments and social and professional idiosyncrasies. As I prefer to deal with the pleasantries of the past rather than its flagrant misdemeanors, it will be perceived that I employ a somewhat gentle phrase in designating a period in the judicial history of the Comstock when judges were corrupted, the verdicts of juries were purchased, and troublesome witnesses were killed or spirited out of the Territory, when mining records were tampered with,

---

[2]Three of the ablest men at the bar were elected district judges—Richard S. Mesick, Richard Rising, and Caleb B. Burbank. Mesick soon went off the bench and was the leader of the bar for years. In the '80s he had a distinguished career in San Francisco. Rising had a long and honorable career as District Judge; incidentally he was said to be one of the most successful poker players in Virginia City. He was a brother of Rev. Franklin D. Rising.

WILLIAM M. STEWART IN 1865

and witnesses before testifying were drilled in perjuries like squads of raw recruits.[3]

The Yellow Jacket claim was located on May 1, 1859, soon after the discovery was made at Gold Hill. There was no ore on the surface so the locators staked the claim over the big quartz croppings on the footwall side of the Lode. Subsequently, the Union and Princess claims were located parallel to and below the Yellow Jacket and the owners discovered ore in a tunnel. The Yellow Jacket then "floated" its claim nearly 300 feet down the hill so as to cover the ore body and brought suit, which was tried in June 1863.

At the trial of the case, which turned upon the early location of the Yellow Jacket and the position of its boundary monuments, 28 witnesses testified for the Princess and Union and 51 for the Yellow Jacket. One Yellow Jacket witness, whose testimony was deemed of great importance, testified to seeing and reading many times the Yellow Jacket notice of location, which he said was posted on a certain stump down the hill. When this witness had concluded his testimony, the attorney for the Princess and Union, General Charles H. S. Williams, dramatically stated that he had but one question to ask the witness. The witness was handed the notice of location and asked to read it to the jury. After an embarrassing pause he broke down and admitted that he could neither read nor write. The cards, however, were stacked and the verdict was in favor of the Yellow Jacket, whose attorneys were William M. Stewart and A. W. Baldwin. During the trial, after the lawyers had wrangled for two days over the location of a certain stump used as a monument, the court and the jury went to view it. The stump had been removed over night and not even the spot where it stood could be identified, so skillfully had the work been done.

In another important mining case, which turned on the position of a location stake, a well-known attorney of professed sanctity is said to have proposed to a witness: "Bill Stewart has paid you a thousand dollars to swear to a lie about the location of that stake. Now, I will give you two thousand to tell the truth." Stewart, who was not thin-skinned, admitted that he "fought fire with fire"; in fact, he appears to have furnished most of the fire. It should be said that the leaders of the bar as a rule were honorable men. When the Comstock declined they scattered all over the West and again stood at the head of their profession.

---

[3]*San Francisco Call*, September 10, 1893.

The most disreputable of all of the blackmailing suits brought that year were those filed in the courts of San Francisco against the Ophir and the Gould & Curry by the "Grosch" Con. Gold & Silver Mining Co., which claimed to own those mines on the preposterous theory that they had been located by the Grosh brothers in 1857, and that the company had acquired the rights of the heirs. The company flooded the West with propaganda, including a faked diary, and sold a large amount of stock privately to a credulous and sympathetic public. The Stock Exchange refused to list it. Among other absurd published statements was the claim that the brothers had sunk a shaft on the Ophir bonanza and extracted several tons of rich silver ore. The Comstock laughed. Shinn thirteen years later swallowed the stuff whole.[4] The Grosch Company postponed the trial of the cases on one pretext and another for nearly two years, and when forced to trial dismissed them without offering any proof whatever.[5] The court required the company to pay the defendants' heavy costs.

Much of the Comstock litigation was disgraceful, and many of the methods employed in the mines and the courts were disreputable, but Lord overdoes it in devoting 60 of 414 pages of his valuable book to that subject.[6] Shinn condenses the story into twelve well-told pages.[7] As a matter of fact, there were but four highly important and long-contested cases: Ophir v. Burning Moscow; Gould & Curry v. North Potosi; Chollar v. Potosi, and Yellow Jacket v. Princess and Union. Three fourths of the others never came to trial.. It is estimated that Comstock litigation during the first five years cost between nine and ten million dollars.

The great copper camp of Butte, Montana, has a record for long-continued and malodorous litigation and skullduggery in the contests between F. Augustus Heinze and the Anaconda Copper interests that makes Comstock methods look like the work of amateurs.[8]

---

[4]*The Story of the Mine*, p. 130 (1896).

[5]*San Francisco Bulletin*, March 9, 1865. See Lord's *Comstock Mining and Miners*, p. 133 (1883). Shinn, p. 130, praised this as "a long and brilliant fight."

[6]*Comstock Mining and Miners*, pp. 97–108, 131–180 (1883).

[7]*The Story of The Mine*, pp. 123–135 (1896).

[8]*Romantic Copper*, by Ira B. Joralemon (1934); Glasscock's *War of the Copper Kings*.

CHAPTER IX

The Comstock Lode—The One-Ledge Theory Prevails—
The Ore Bodies.

The Comstock Lode was formed in a complicated fault fissure which extended along the base of the Virginia Range for a distance of about four miles and reached to unknown depths. Ages passed before the Lode reached completion. The process began during the violent Tertiary period, when the east face of the range was covered by repeated flows of igneous rock from volcanic vents a few miles to the eastward. The region was subjected to violent dynamic convulsions and a great fault fissure was opened on the line of contact between the overlying volcanic flows and the east slope of the range. Von Richthofen was of the opinion that the fault was a normal one—that the upper or hanging wall side slid downward on the footwall. Becker inclined to the view that the range was shoved upward, although he states it both ways.

In the course of time the upper volcanic flows were eroded almost to the present level and repeated dynamic movements reopened the fault fissure, shattering the walls and leaving more or less open spaces between them which extended to subterranean depths, affording escape for volumes of mineral-bearing hot waters, steam and gases, "which, bursting from a hundred vents, rapidly decomposed the surrounding rocks and gradually filled the fissures of the Comstock with their remarkable charges of mineral-bearing quartz."[1] The Comstock during that period was Steamboat Springs magnified a hundred times.[2]

---

[1] Clarence King in Vol. 3 of *U. S. Exploration of the 40th Parallel*, pp. 95, 96 (1870).

The first report on the geology of the Comstock, written by Baron Ferdinand Von Richthofen in 1865, has been esteemed a work of genius. His forecast of the lode in the light of subsequent developments is almost uncanny. The report was printed in San Francisco on November 22, 1865, and reprinted in 1866 and 1868 by the Sutro Tunnel Company.

Von Richthofen was followed by Clarence King, who wrote the preliminary chapter on geology in James D. Hague's notable volume on the Comstock, *U. S. Exploration of the 40th Parallel*, Vol. 3 (1870). The third important report was George F. Becker's *Geology of the Comstock Lode*, published by the U. S. Geological Survey in 1882. Becker, in a preliminary chapter, pages 12 to 31, narrates briefly the conclusions of Von Richthofen and King and those of Professor John A. Church, who wrote a report in 1877. The several later minor

When the fissures of the Comstock had been filled with quartz and sealed at the surface and the diminished deep-seated hot waters could no longer find an outlet, they never ceased to flood the lower mine workings and became the greatest obstacle to mining.   On the deep levels the hot water all but stopped development work.

The east slope of the Virginia Range, which extends downward indefinitely at an average dip of 45 degrees, forms the footwall of the Lode, which follows the contours of the range.   In the Virginia City section it is composed of hard granitoid diorite. At Gold Hill the footwall is metamorphosed slates.   The hanging wall is a highly altered andesite, which the miners called porphyry.[3]

It is generally agreed that the first great masses of quartz which formed in the Lode were so poor in silver and gold as to be practically valueless, and that the rich ore bodies were deposited in the low-grade quartz at a later period by highly mineralized solutions which found entry when the fissures were reopened from time to time.   The quartz after its deposition was repeatedly subjected to dynamic pressure and extensively fractured.   The rich ore was nearly always soft and easily mined, much of it having been crushed by movement "until it had the appearance of crushed sugar."[4]   The pressure continued at intervals after the violent movements which shattered the great fault. Clarence King wrote that the extensive sheets of clay in the Lode were caused by the rapid decomposition of the propylite by solfataric action.   There were no later important dislocations; the Lode was not faulted nor were there any displacements of the ore

reports are cited on pp. 31, 32 of Vincent P. Gianella's *Geology of the Silver City District*, University of Nevada Bulletin, Vol. 30, No. 9 (1936).

[2]Steamboat Springs described in Becker's *Geology of Quicksilver Deposits*, pp. 331–353 (1888).   The formation was like that of the Comstock.

[3]The diorite was called syenite by Von Richthofen, and the altered andesites within the Lode he called propylite.   Later geologists, after more detailed studies, have agreed upon the terms used in the text.

[4]Becker's *Geology of the Comstock Lode*, pp. 17, 270.   He maintained that the precious metals of the Lode were derived from the diabase hanging wall by lateral secretion, Chap. XV, pp. 225, 385, 396; contrary to Von Richthofen, King, Hague, and other geologists, who assumed that "the vein was filled from a deep-seated source."

Edson S. Bastin, who made an examination of the Comstock in 1913, said that Becker's lateral secretion theory has been generally abandoned. *Bonanza Ores*, p. 45 (1922), *U. S. Geological Survey Bulletin*, 735C.

King said that the "black dike," a thin sheet of hornblende andesite lying upon or near the footwall, "was undoubtedly the first step in the formation of the Lode."

bodies, except perhaps at Gold Hill. There the geologists were divided upon the question whether the west vein was a faulted segment of the east ore bodies.

A remarkable feature of the Lode is that the wide upper section appears to have been formed in a V-shaped trough, from 200 to 1,000 feet wide at the surface, which narrows rapidly to the depth of 500 feet or more where the Lode assumes an average width of 100 feet and dips eastward. This V-shaped section is filled mainly with large blocks of porphyry, which were split off the overhanging east wall in long sections by a series of nearly vertical faults while the Lode was being formed. It was the splitting off of these great fragments and their inclusion in the Lode that gave it the extreme width at the outcrop, and gave to the hanging wall ore bodies a westerly dip near the surface. These masses of porphyry, called "horses" by the miners, were up to 1,000 feet in length and hundreds of feet in width at the surface, narrowing like wedges at the bottom. They were often reduced to spongy masses more or less filled with quartz and clay. Quartz and sheets of clay formed within and on both sides of the "horses," but the upper ore bodies were formed in the nearly vertical fissures, which at first appeared to be separate veins. A "horse" in a wide lode may be likened to an island in a frozen river.

Von Richthofen said that five sixths of the Lode in the upper portion consisted of "horses," the remaining one sixth being made up of quartz and clay. Clarence King, in 1868, estimated that not to exceed 1/500th part of the Comstock Lode was profitable ore.

The ore itself was always more or less filled with fragments of porphyry, varying in size from small grains to bodies of considerable size, which were often sufficiently mineralized to be called ore. Old-time Comstock miners had a saying "porphyry makes ore." Becker stated that "The evidence appears conclusive that Comstock ore bodies occupy spaces which once enclosed only fragments of country rock, with numerous interstices."[5] The so-called "underground reservoirs" in the upper part of the Lode occupied similar spaces and were retained by walls of clay. This was all surface water. The uprising hot water in the deep levels was never fully overcome.

---

[5] *Geology of the Comstock Lode*, p. 395 (1882).

Becker made experiments to determine the electrical activity of the ore bodies, thereby anticipating modern geophysical prospecting. Id., pp. 400–404.

THE ONE-LEDGE THEORY PREVAILS

John Nugent,[6] had been appointed by Judge North as referee to hear the testimony in the case of the Gould & Curry against the North Potosi and report to the court. The report, which dealt solely with the geological features of the Lode, was masterly and convincing and formed the basis of the court's decision. The hearing before the referee had consumed weeks. Dozens of witnesses had given their testimony—among them such notable experts as William Ashburner, Prof. Blake, Dr. John Veatch, I. E. James, Capt. Sam Curtis, Prof. Phillips, Prof. Benj. Silliman, Prof. Willis, and Prof. Whitney.

The North Potosi claim had been located on an assumed "blind ledge" and lay east of and adjoining the Gould & Curry, Savage, and Hale & Norcross companies' claims. No friction arose until the Potosi company ran a tunnel and encountered a vein of rich ore in a fissure in the hanging wall lying east of the Gould & Curry bonanza. The Potosi claimed that it was a separate vein, while the Gould & Curry maintained that it was merely one of the many vertical fissures in the hanging wall, separated from the Gould & Curry fissure by a wall of clay.

The referee's report, dated August 21, 1864, found that the Comstock was a true fissure vein, lying between two formations of entirely different origin and character; on the west a hard tough footwall rock of syenite, dipping eastward at an average angle of 45 degrees, and on the east a soft and friable mass of feldspathic porphyry containing many slips and fissures. He found that there was no true east or hanging wall in the wide upper section of the Lode, but a succession of splits or fissures in the east porphyry extending upward from the Lode, always carrying clay and sometimes ore, and that the North Potosi fissure was of that origin and formed a part of the Gould & Curry claims.

The referee found that the large "horses" in the Lode were originally part of the overhanging east wall, from which they were broken off and fell into the great Comstock fissure, where they became surrounded by quartz and clay.[7]

---

[6]Nugent had been a prominent lawyer in San Francisco and was one of the ablest men at the Nevada bar.

[7]The report was printed in the Virginia City and San Francisco newspapers and was issued later in pamphlet form. Lord discusses this hearing and the referee's report on pp. 165–171 of *Comstock Mining and Miners*.

THE ORE BODIES

The ore bodies were thinly scattered through the wide Lode "like plums in a charity pudding," as Mackay expressed it, and nearly all of them were found in the wide upper section and along or near the east wall. The only great ore body lying on the normal footwall was the Crown Point-Belcher bonanza, which extended on the dip of the Lode from the 1,000 to the 1,600-foot levels. The Con. Virginia bonanza stood in a vertical shattered rent in the hanging wall, which broke upward from the Lode at what became the 1,750-foot level, and ceased 1,100 feet vertically below the surface. That fissure was not known to exist until it was encountered accidentally on the 1,200-foot level, where it stood 700 feet east of the Lode. The Lode in that section had contracted to a mere fissure below the 900-foot level. Von Richthofen, in his remarkable forecast of the future of the Lode in 1865, when the Comstock was deeply depressed, anticipated the Con. Virginia bonanza in the statement:

It is probable that repeatedly, in following the Lode downward, branches will be found rising from its main body vertically into the hanging wall and consisting of clay and quartz. Many of them will probably be ore-bearing.[8]

Other fissures did arise from the lode far to the eastward, but as far as known the Con. Virginia fissure was the only one containing profitable ore. He ventured the further prophecy: "Some mines which have been heretofore almost unproductive, as the Central, California, Bullion, and others have therefore good chances of becoming metalliferous in depth." That guess was good only in part. Neither he nor any other geologist could have anticipated that the Lode would continue barren for such long distances and to such great depths.

While Von Richthofen heartened the miners with the assurance that the Comstock was a true fissure vein and that other large ore bodies would be found far below, he was in error in assuming the ore would be of lower grade and that the recurrence of rich bonanzas like the Ophir and the Gould & Curry must not be expected. The Crown Point and Con. Virginia bonanzas were giants by comparison. They produced one half of the total production of the Lode and paid four fifths of all the dividends.

---

[8]Quoted by Becker, p. 24, *Geology of the Comstock Lode.*

There was no guide in the search for an ore body in that wide Lode. The miners simply extended their workings in every direction on level after level in the hope of success. A vain search it was on the whole, for only sixteen large and rich ore bodies were ever found in the Comstock—and most of these lay within 600 feet of the surface. The rich ore had a way of concentrating in large bodies with the highest grade in the center. Very few small deposits of ore were found except in association with large concentrations. As a rule the ore bodies occurred somewhat in the form of lenses, sheathed in walls of clay and accompanied by parallel sheets of ore separated from the main body by porphyry walls of varying thickness and by other sheets of the abundant clay.[9] Even when the ore shoots were of considerable length horizontally they were usually made up of parallel or coterminous lenses.[10]

After the discovery on the surface of the Ophir and the Original Gold Hill bonanzas in 1859 the remaining ore bodies along the Lode were found from time to time by trial and error over a long course of years, thereby lengthening the life of the camp. None of them lasted over five years. As a rule they were "gutted" of rich ore to make a play in the stock market.

The wide Lode in the upper section was a puzzle to the early miners, and several years passed before they learned to look for ore in the nearly vertical fissures along the so-called east wall. Even then they were slow to learn that other lenses or sheets of ore might be found beyond what appeared to be the wall. One reason was that they always hesitated to cut through a clay wall for fear of tapping a deluge of water.[11] Much of the early exploration work was done in the wide belts of almost barren quartz lying on the west side of the great "horses" that filled the Lode, and it was not until 1865 and 1866 that they learned to drive long crosscuts easterly through the "horses" to the ore bodies lying along or near the east wall.

The principal silver values in Comstock ores were in the sulphides — argentite, polybasite, and stephanite. Native silver

---

[9]Excellent plates illustrating the lode and the various ore bodies are printed in the separate atlases which accompany Hague's Vol. 3 of *U. S. Exploration of the 40 Parallel* (1870), and Becker's *Geology of the Comstock Lode*, printed by the U. S. Geological Survey in 1882.

[10]The series of early ore bodies in Virginia City and at Gold Hill extended horizontally as a rule. They were formed near the surface and spread north and south within the walls of the Lode.

[11]On January 20, 1863, Ophir miners on the 300-foot level tapped an underground reservoir and had to run for their lives.

occurred sparingly. The gold was free. The ore was always associated with iron pyrite, and in the Ophir and the Gould & Curry bonanzas some lead, zinc, antimony, and copper sulphides were always present. The ores in the Gold Hill section in contrast were practically free from base metals.

The great Lode splits into two branches at each end—at the Ophir on the north and at the Belcher on the south. One of the forks at the north end extends onward to the Utah, the other turning northeasterly to the Scorpion. Similarly, at the south end, one fork extends beyond the Belcher to the Baltimore, while the other turns southeast down Gold Cañon beyond Silver City. It is worthy of note that all of the great ore bodies were found in the main portion of the Lode, within a distance of two miles. The large Silver City branch produced considerable low-grade ore but was profitable in the early days only in small part, although several millions were spent in development work.[12] The Utah and Scorpion branches were practically valueless.

There were many parallel veins on both sides of the Comstock and distinct from it. None of them proved profitable, although millions were spent upon them over a long course of years. The most important parallel vein was the Occidental, which extends across the hills nearly two miles east of the Comstock. It was extensively mined, off and on, and produced about $500,000, but at no profit except during a brief period in the early days. That vein has the same eastward dip as the Comstock Lode.

Cedar Hill, above the Comstock Lode, is peculiar in that it is traversed by many little gold veins upon which a great deal of fruitless development work was done. The debris at the foot of the Hill on the Sierra Nevada ground was thought rich enough for hydraulic operations. Two such attempts were made, one in 1862 and the other in 1874, both of which were failures.

The accompanying map, showing the course of the Lode through the mines, was made in 1880, but the names and the boundaries of nearly all of them are represented as they were in the early '60s. Then and later several consolidations were made.

The Sierra Nevada had acquired about fifty mines on Cedar

---

[12]The Justice, in Gold Cañon above Devil's Gate, found a substantial body of ore in 1875, which should have been profitable. It produced 148,319 tons, yielding $2,759,356, or $26 a ton. Instead of dividends, the management continued to levy assessments. George Schultz, the president, was severely criticized for questionable practices by an investigating committee and deprived of control. *U. S. Mineral Resources for 1872* p. 115, refers to a large area bearing low-grade gold ore below Gold Hill which may become important.

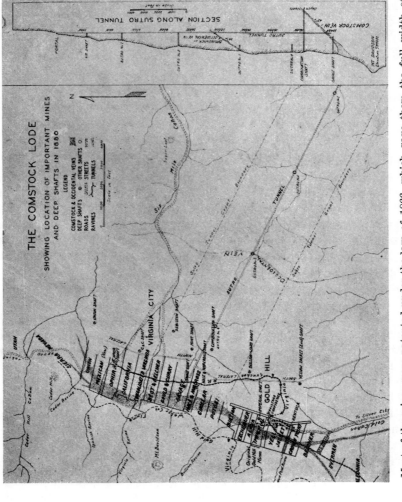

Most of the mines were patented under the law of 1866, which gave them the full width of the Lode, and did not require side lines. From Atlas Sheet accompanying Becker's Geology of the Comstock Lode.

Hill, some of which were thought of great importance in the early days but had no real value.

The Consolidated Virginia and the California were formed on six little mines, which failed to find ore after $1,000,000 had been spent upon them.

The Mexican shown on the map was a new company, incorporated in 1874, which was given the north 600 feet of the Ophir. The old Mexican was absorbed by the Ophir in 1867.

The Chollar and the Potosi, which were consolidated in 1865 after scandalous litigation, were segregated into separate mines in the late '70s.

The rich Little Gold Hill Mines were incorporated as the Cons. Imperial after their ore had been worked out.

All of the other mines had either absorbed conflicting claims or had defeated them in litigation.

Sutro did not begin to dig his tunnel until 1869 and it did not reach the Lode until nine years later.

The upper streets of Virginia City and Gold Hill were laid along the Lode.

CHAPTER X

The Early Bonanzas—The Ophir, The Gould & Curry, The Savage, The Chollar-Potosi, The Yellow Jacket, and The Original Gold Hill Bonanzas.

THE OPHIR BONANZA

The fame and the prosperity of the Comstock during the first five years were based upon six bonanzas of which the Ophir was the most famous.

The notorious extravagance of the Ophir management during the early years was not due to lack of able men in its affairs.[1] They were free-handed Californians, glorying in the possession of the greatest silver mine in the world, as they believed. The board of trustees, relying upon the traditional inexhaustibility of silver mines, built the Ophir grade down to Washoe in 1860, and erected a great mill there the following year. The location was chosen because of the abundance of wood and water. As the ore was complex the Bagley variation of the Freiberg process (which involved dry-crushing by stamps, chloridizing-roasting, and amalgamation in revolving barrels) was thought the only one suitable.[2] Bagley guaranteed that it would cost less than $30 a ton to mill the ore by this process. Very soon, however, a "Washoe process" mill was added to reduce second-class ore, also patio yards for low-grade ore.

The mill and the surrounding buildings were built without regard to expense; a number of the structures were of cut granite. The total cost was not less than $500,000. In order to save a three-mile haul around the lower end of Washoe Lake the wagon road was carried straight across the lake for a mile and a half on a piling bridge. A railroad was even planned and surveyed from the mine to the mill.

The mill was fairly successful while the rich ore lasted, although the cost of reduction by the Bagley process considerably exceeded $50 a ton. The first 3,000 tons of selected ore milled in 1862 yielded $326 a ton, after losing not less than one third of the values in the tailings.

---

[1]The company was organized April 28, 1860, with a capital of $5,040,000, divided into 16,800 shares. The incorporators were Wm. M. Lent, William Blanding, J. H. Atchison, Squire P. Dewey, Edward Stanley, John O. Earl,

The managers and the stockholders of the Ophir walked on air during the first two years while the mill was building and the bonanza continued to improve. Then troubles began to multiply: the mill was more or less of a failure from the first; the company became involved in thirty-seven different lawsuits with adjoining mines; and the ore body gradually narrowed in length and decreased in size and value below the 300-foot level until it terminated in a thin wedge of low-grade material at the depth of 500 feet.[3] Not only did the ore become thinner and poorer but a large "horse" of worthless porphyry nearly filled the ore body toward the bottom. An unexpected benefit from the caving of the Mexican from top to bottom on July 15, 1863, was the exposure of a parallel sheet of rich ore standing 40 feet west of the main body. It was 300 feet long, from 5 to 15 feet wide, and extended from the surface to the 350-foot level. That was the last high-grade ore found in the Ophir and the Mexican and was worked out during 1864. The big Ophir mill ceased to operate in 1866, and its site is now nearly obliterated; even the granite blocks have been carried away by the neighboring farmers. The Ophir had but three prosperous years—1862, 1863, and 1864.

The litigation with the Burning Moscow Company, which had the large footwall vein lying west of the Ophir, was savagely contested for three years, with the fortunes of war generally favoring the former. The stocks of both companies fluctuated widely from time to time as advantages shifted; armed men met in conflict in the lower workings on several occasions, and every device and expedient that ingenious lawyers could suggest was employed.[4] Finally, in the midst of the depression of 1865, which

---

Wm. L. Dall, Charles F. Lott, Louis A. Garnett, R. L. Ogden, William Thompson, Jr., Theodore Winters, Jos. C. Vandervoort, Wm. F. Dewey, and Joseph Woodworth. The board of trustees was composed of Wm. F. Babcock, Joseph Woodworth, Chas. F. Lott, Wm. M. Lent, Squire P. Dewey, and Theodore Winters. William Blanding was made president, James W. White secretary, William C. Ralston treasurer, and Wm. L. Dall (an old sea captain) superintendent. Later such men as J. B. Haggin, Jesse Holladay, Edward Martin, J. M. Livingston, and Donald Davidson served as members of the board. A number of those men became wealthy and powerful in California, but few of them profited in the Ophir.

[2]The Ophir mill and the processes are described in detail by Almarin B. Paul in the *San Francisco Bulletin* of October 24, 30, 1862.

[3]Wm. Ashburner, California State Mineralogist, wrote in November 1865 that the Ophir shaft was 606 feet deep, and that no ore had been found below the 7th level, which was 420 feet below the surface. Samuel Bowles' *Across the Continent*, p. 448 (1866).

[4]Lord gives a stirring account of that litigation and accompanying events, *Comstock Mining and Miners*, pp. 138–144 (1883).

proved fatal to litigation all along the Lode, the contest fizzled out and the Ophir purchased the Moscow property for $70,000. It is said that the controversy cost the two companies about $1,000,000.

The Ophir bonanza was rich but comparatively small. It stood in a fissure next to the hanging wall and sloped to the west. The ore body was 500 feet in length along the surface. Beyond the ore at each end the vein was filled with base low-grade ore for some distance which was unprofitable when worked in later years. The total production, including the portions mined by the Mexican and the Central, did not exceed $7,000,000.[5] The yield of the Ophir was 70,000 tons, of an average value of $75 a ton, or $5,250,000, out of which $1,394,000 was paid in dividends, the last in March 1864. Ten years elapsed before it found another ore body and never another like the first. The company began to sink its third line shaft on "E" Street in 1867, in which it had an endless battle with water.

In 1874 a small ore body was discovered on the 1435-foot level adjoining the Con. Virginia-California bonanza. It extended to the 1700 level and yielded $4,351,492, coin value, from which no dividends were paid under Sharon's management. In 1877, after the Bonanza firm had taken control, a narrow east-west vein was found on the 1900 level, from which two dividends of $108,000 each were paid.

From 1904 to 1911, after the north-end mines had been pumped out to the 2500-foot level (under the administration of the few remaining S. F. brokers), several small, rich ore bodies were found trailing from the 1900- to the 2400-foot level, which yielded $3,001,837, but brought only $262,080 in dividends. Below that, to the 3300 level, no ore was found.

The control of the Ophir passed into the hands of successive groups of speculators and the stock was one of the most active on the market for many years. In all its history the mine did not create a single millionaire. E. J. Baldwin, who was already rich, was wise as well as "Lucky" when he sold 20,000 shares to Sharon for $2,700,000 in November 1874 during the Con. Virginia boom.[6]

The mine was a continuing disappointment throughout its long career. The total production from 1859 to 1939 was $15,600,000;

[5]Vol. 3, p. 169, *U. S. Exploration of the 40th Parallel* (1870) by James D. Hague; Ophir annual reports.

[6]*Virginia City Chronicle*, December 12, 1874.

total dividends $1,872,480; total assessments $5,600,000, a large part of which was expended for deep mining. Water was the bane of the Ophir for years.

THE MEXICAN, A PART OF THE OPHIR BONANZA

The successive owners of the Mexican had nothing but hard luck. During 1860 and 1861 Maldonado borrowed $170,000 from Alsop & Co., of San Francisco, to build an elaborate Freiberg mill. He was extravagant and unable to pay, and, after a lawsuit and a compromise the mine was taken over by Alsop & Co. The mill burned in April 1863, two disastrous caves occurred later in the year, and the mine could not provide ore for the large and costly mill erected at Empire on the Carson River. The narrowing of the wedge-shaped Ophir bonanza cut out the Mexican ore at the depth of 300 feet, although the discovery of the narrow vein of rich ore paralleling the main Ophir ore body on the west provided much-needed income during 1864.

The mine was worked by Mexican miners after the fashion of their country during the first two years. Instead of hoisting the ore they packed it up the incline shaft in rawhide buckets slung from the forehead. For ladders they used notched logs.

Although the Ophir began to install square-set timbering in the fall of 1860, the Mexicans left pillars of ore in the stopes to hold up the roof. These began to crumble in time, and, despite the efforts of later American miners to remedy the situation, the mine caved from the surface to the 225-foot level on July 15, 1863, completely wrecking it and damaging the Ophir workings. A second cave occurred six months later, after which the old workings were abandoned and a new vertical shaft was sunk a short distance eastward in conjunction with the Ophir. The new joint Ophir-Mexican shaft produced little except floods of water, and the mine was absorbed by the Ophir in 1867. The old Mexican did not produce to exceed $1,000,000, and never paid a dividend. The stock was not listed on the exchange.

In 1874, during the boom, "Lucky" Baldwin, who was in control of the Ophir, formed a new corporation called the Mexican, which was given 600 feet off the north end of the Ophir. The stock was a favorite gamble for years, but paid only $161,190 in dividends (in 1912, 1913, after the north-end mines had been pumped out) as against $4,000,000 in assessments in all its history. The mine was worked through the Ophir main shaft.

THE GOULD & CURRY BONANZA

The mine was a consolidation of the Gould claim located by Alva Gould, and the Curry claim located by "Old Abe" Curry.[7] Curry had located on the prominent El Dorado croppings, the highest point on the Comstock Lode. Gould took the ground below him on the mountainside. Both claims soon passed into the hands of twelve men, chiefly San Franciscans, who organized the Gould & Curry Silver Mining Company on June 25, 1860, with 4,800 shares, four to each of the 1,200 feet of the Lode which they claimed at the time. Among the incorporators and trustees were George Hearst, Lloyd Tevis, John O. Earl, Alpheus Bull, Thomas Bell, A. E. Head, B. F. Sherwood, and William Blanding, all of whom became California millionaires after making a good beginning in the Gould & Curry. John O. Earl was made president, William C. Ralston treasurer, and Charles L. Strong superintendent. Strong arrived in the fall of 1859 as agent for Wells Fargo & Co.

The property had been purchased on the strength of the large croppings and a little surface ore, and was operated in a small way during 1860 and 1861.[8] The footwall El Dorado croppings proved barren, but a narrow vein of good milling ore was found in a shallow tunnel 1,000 feet down the slope on the east side of the Lode, which gave little promise of the great bonanza 100 feet beneath. It was not until the end of 1861, when the "D" Street tunnel penetrated 40 feet of rich solid ore that the mine began to overshadow the Ophir and arouse hopes of future greatness that turned to ashes in the course of a few years. Instead of continuing downward the ore body pitched southward into the Savage where it was equally productive. Neither the Gould & Curry nor the Savage found another ore body in all its history,[9] with the exception of those which the Savage shared with the Hale & Norcross from 1866 to 1869.

---

[7]DeGroot said Curry received six or seven thousand for his claim, and Gould even less.

Abraham Curry was one of the foremost citizens of Carson City for many years. Mark Twain praises his public spirit in chapter 25 of *Roughing It*.

[8]The records of the trustees show that in 1860 they borrowed $10,000 of R. B. Woodward (the proprietor of the What Cheer House in San Francisco and of the famous Woodward's Gardens) at three percent per month, and that during 1861 assessments were levied to the amount of $166,068. The production in 1860 was $22,004, and $44,221 in 1861. Its greatest years were 1863 and 1864 when the yield was almost $8,000,000.

[9]The steep slope of the mountain below the Gould & Curry croppings favored development by tunnels. The lowest tunnel, which was 280 feet below the one on "D" Street, found the ore body smaller and poor in grade. Below that the ore gradually petered out.

In 1861 the Gould & Curry began to build a small mill, but, after the discovery of the bonanza, the trustees of the company, believing that the ore would last indefinitely, ordered the erection of that monumental mill in Six Mile Cañon, a mile and a half below the mine.[10]   As the ore was rich and complex, the Dr. John Veatch variation of the Freiberg process was adopted. His preliminary tests showed that the ore could be milled for $15 a ton, with an 84 to 91 percent recovery, but when the mill went into operation in 1863 the cost of reducing 4,892 tons of high-grade ore was "from $50 to $60 a ton," and the loss in the tailings was said to be 40 percent.[11]   At that, the ore yielded $316 a ton.   In 1863, 48,743 tons milled yielded an average of $80.44 a ton.

Fortunately for the stockholders the greater part of the production of the mine during 1863 and 1864 was reduced in custom mills employing the Washoe process, at a cost of $26 a ton, with a 65 percent recovery.

The extravagant spirit of the times is illustrated by the fact that the management and the general public took pride in the mill until it proved a failure.   It was a showy period; wealth was expected to make a display.

During 1864 the mill was practically rebuilt in order to employ the Washoe process, under the direction of the new superintendent, Louis Janin, Jr.   The number of stamps was increased from 40 to 80, and 40 pans and 20 settlers were installed.   The total cost of the mill and its accessories, including the expense of remodeling, was $1,500,000.[12]   The mine could then provide only ore of moderate grade, which Janin succeeded in milling at a cost of $13 a ton, not counting capital investment.   By the end of 1866 the mine was practically out of ore and the mill was closed down—to stand for years a decaying monument to Comstock prodigality.   The ore body was mined over a length of 500 feet and a width of 100 feet, at its best, including parallel sheets and stringers, and at the depth of 500 feet passed entirely into

---

[10]While the mill and surrounding buildings were elaborate, Lord's description in *Comstock Mining and Miners* (pp. 124, 125), is overcolored, as appears by the photographs of paintings made in 1863. The mill and the process were described in detail by A. B. Paul in an article published in the *San Francisco Bulletin* of October 22, 1863.

[11]Prof. Benj. Silliman wrote in his report on the Potosi, dated June 24, 1864, that the loss of the Gould & Curry mill in the tailings was probably 40 percent, and that the slimes in the slime pit assayed above $130 a ton.

[12]The failure of the big mill in 1863 and the dissatisfaction with the management, brought a change of officers in 1864.   Supt. Charles L. Strong was supplanted by Charles Bonner.   The *Enterprise* of November 7, 1878, says that Strong is managing the White & Shiloh mine in Lander County.

the Savage. The northern 500 feet of the mine was barren. No
ore was ever found below 500 feet, although a small body of low-
grade ore went 50 feet deeper.

During the years 1862, 1863, and 1864 the company shipped
50½ tons of ore to Wales for treatment, which averaged $1,800
a ton.

Gould & Curry ore averaged $70 a ton during the first four
years, then decreased in value. The total production was 315,000
tons, yielding $15,750,000, or $50 a ton. Dividends to the amount
of $3,826,000 were paid to and including 1865, followed by an
unwarranted trifle of $40,000 in 1870 as a stock market stimulant.
The assessments levied to 1939 amount to $6,780,000, most of
which was spent in fruitless deep mining.

The stock, which had sold at $75 a foot in December 1860,
reached its highest point, $6,300 a foot at the end of June 1863,
then gradually declined.[13] Meantime, some of the well-informed
insiders unloaded their stock on the public. Comstock mining was
always carried on with an eye on the stock market.

THE SAVAGE

The Savage was a child of fortune. Its first great ore body was
found for it by the Gould & Curry, and the next, four years later,
by the Hale & Norcross, its neighbor on the other side.

The Savage had spent little on the mine except for litigation
when the Gould & Curry bonanza was discovered. That ore lay
in the south end of the claim, and shrewd and aggressive Robert
"Bob" Morrow, superintendent of the Savage, obtained permission from the Gould & Curry to drive a drift southward from the
latter's "D" Street workings into the Savage. Just as Morrow
expected, the bonanza extended southward into the Savage.

The Savage then began to sink a new shaft in order to extract
its ore and develop the mine and did not commence to produce
until April 1863. For the next two years it was in bonanza.
The yield was 81,183 tons, averaging $44.35 a ton, a total of
$3,600,709, almost one half of which was paid for milling. The
ore was not as high grade as in the Gould & Curry, although in
one stope "it had a width of 90 feet of rich solid ore"; nor did
the ore contain so much base metal. It was reduced in Washoe
process mills, with a return of 65 percent of the assay value. The
company built a mill in Washoe Valley, and bought another there

---

[13]In 1874 the Gould & Curry conveyed 321 feet of the north end of the mine
to the Best & Belcher for stock in the latter.

in 1865—failing to make allowance for winter weather, like other Washoe millmen, which prevented the hauling of ore over the Virginia range for four months of each year. At the height of the "Boom of 1863" the leading stockholders sold out and wisely invested in San Francisco real estate. Capt. Sam Curtis, who was given to extravagance, then became superintendent and continued until July 1866, when Charles Bonner took charge. At the end of 1865 the mine was on the decline. It had paid only $800,000 in dividends, was in debt $417,237, and the rich ore appeared to have given out. Further development work, however, soon disclosed the extension of the ore body and the discovery of the rich little "Potosi strike," which lay in an easterly split in the hanging wall. Another stroke of fortune was the discovery by the Hale & Norcross of a fine ore body on the 600-foot level in December 1865, one half of which proved to be in Savage ground.

Then followed three years of large production, during which the Savage was the great mine of the Comstock, with a yield of almost $8,000,000 and the payment of $3,408,000 in dividends, the last in 1869. That ended the prosperity of the mine. The ore in the Savage and the Hale & Norcross occurred in several parallel bodies. The deepest went to the 1400-foot level, but the ore was base and low grade below the 1200.

The Savage struck a flow of hot water in 1876 on the 2200-foot level that flooded the mine and the Hale & Norcross to the 1800-foot level. Both mines remained flooded for three years, despite continual pumping, until a connection was made with the Combination Shaft on the 2000-foot level in 1879. Assessments commenced in 1870 and never ceased. They averaged $400,000 a year while the Combination Shaft was in operation. Floods of hot water prevented connection with the Combination Shaft below the 2400. The levels below that had to be bulkheaded.

In 1871, after the discovery of the Crown Point bonanza, John P. Jones and Alvinza Hayward took over the control of the Savage and ran the mine for the benefit of their mills. In the annual report of July 1, 1872, the superintendent, genial A. C. "Lon" Hamilton, stated that 47,505 tons of ore had been mined and milled at a loss of $3.25 per ton. The next year 33,414 tons were milled at a loss of $13.29 per ton, and assessments amounting to $640,000 were levied.

Hayward began to boom Savage stock early in 1872 by giving out mysterious reports of a rich strike, and by confining the miners underground—an old Comstock trick. The miners did no work and lived on the fat of the land.   Great excitement followed. The price of the stock rose rapidly from $62 a share on February 1 to $725 on April 25.   In May the Crown Point boom collapsed and Savage with it.   There had been no rich strike; it was a cold-blooded stock deal that hurt many people.[14]

THE CHOLLAR–POTOSI

The Chollar-Potosi was a consolidation of the Chollar and the Potosi claims, effected in 1865, after four years of litigation that shook the Comstock to its foundations.

The Chollar was located in 1859, 1,400 feet in length and 400 in width, on the low-grade footwall croppings; the Potosi was located, adjoining the Chollar on the east, also claiming 1,400 feet in length and 400 in width.

Neither had any ore until the latter part of 1861 when the Potosi discovered an ore body dipping into the Chollar.   The latter brought suit, by its attorney Wm. M. Stewart, and the fight was on.   After two trials the Chollar won a verdict and the decision was affirmed by the Supreme Court.

The Chollar then found an ore body that dipped into the Potosi, which, in turn, brought action.   The record is not clear, but it seems that the first Chollar decision fixed its easterly side line vertically.   In the Potosi case Stewart tried desperately to reopen the earlier Chollar case by joining the Grass Valley claim, lying east of the Potosi, which the Chollar had acquired.

The courts ruled against Stewart and he was outraged, as he always was when anything went against him.   Meantime A. W. "Sandy" Baldwin had become his partner.   Stewart was an indomitable and ruthless fighter, and in this case he had cause— at least in the conduct of some of the judges.

That controversy, which resulted in the resignation of three judges of the Supreme Court, is told at length by Lord,[15] who obtained from Stewart much of his information and his "color" concerning Comstock litigation.   Lord's pages should be read with that in mind.

The Potosi, located unknowingly over the hanging wall ore bodies, ran a shallow tunnel 218 feet below "datum point" and crossed a solid body of quartz for a width of 280 feet;[16] the easterly 80 feet of which was good ore, while the other quartz was

[14]The *San Francisco Chronicle* of May 19 charges Jones and Hayward with unloading Savage on their friends.

[15]*Comstock Mining and Miners*, pp. 151–163 (1883).

[16]Clarence King in Hague's memoir, Vol. 3 of *U. S. Survey of 40th Parallel*,

blocky and of lower grade. Above that level the quartz divided into three large forks and spread out fanwise to or near the surface, which accounts for the masses of quartz found later throughout the property. The richest ore continued upon or near the hanging wall, but not less than twelve profitable ore bodies were found over a length of 1,000 feet and a width of 300 or more. The hanging wall curved around to the west and south above the tunnel level, and there made the productive Santa Fe and Blue Wing stopes, and, greatest of all, the Belvidere. The latter, 140 feet in length and 45 in width, extended upward at a slight angle and came within 300 feet of the north end of the Bullion. That shoot of ore, which was found when the ore in the mine was almost exhausted, produced $4,000,000 in $40 ore and paid $2,000,000 in dividends, which exceeded all of the dividends theretofore paid.

The official reports begin with the consolidation, according to Comstock practice, and disregarded earlier production by the Potosi and the Chollar, amounting to not less than $2,500,000. Early assessments and dividends are excluded also. Prof. Benj. Silliman, Jr., of Yale, who made an examination of the Potosi in May and June 1864, stated that the mine produced 20,673 tons of ore from May 1, 1863, to May 30, 1864, assaying $75 a ton and yielding $50. The custom mill charged $25 a ton, so the company actually received only $25 out of $75 ore. He advised that it build its own mill which was never done. Sharon obtained control in 1867 and his mills handled all of the ore thereafter. Isaac L. Requa, a competent man and a rare character, was superintendent for twenty years, from 1867 to 1887. The ore averaged only $24.50 a ton from 1866 to 1870, when the Belvidere brought it up to $41.30. The following year it fell to $26.17, then lower still. In 1878, when production ceased,[17] the mine had yielded $16,399,600, paid $3,579,925 in dividends, and levied assessments totaling $2,317,500. These figures including those prior to consolidation in 1865.

Several of the stopes were just below the surface and when the timbering rotted two large caves occurred.

Roswell K. Colcord[18] told of the first cave, which occurred in

p. 45. He says that the walls of the Lode came together at 600 to 800 feet. "Datum point," on the Gould & Curry croppings, was the point from which all official levels were calculated.

[17]Much low grade unprofitable ore was mined from 1887 to 1894, and, after 1920 over 1,000,000 tons were mined at and near the surface—all at a loss.

[18]Ex-Governor R. K. Colcord was the grand old man of Nevada. He celebrated his ninety-eighth birthday on April 25, 1937, vigorous in mind and body,

June 1867. He was then superintendent of the Imperial-Empire shaft at Gold Hill, and he and a friend had walked over to Virginia City to attend the theater. On the way back they were joined by a young fellow who said he was a clerk in Wood & Goe's store and slept there. When they reached the intersection of "B" and "C" streets where the brick store had stood, they saw a great smoking cavern in which the store had disappeared. The young fellow was thunderstruck. When he could talk he said that he had been spending the evening in a hurdy-gurdy house, learning to dance. If that was sinful, he said, he had rather be a sinner than buried in that hole.

The Gold Hill News told of the second cave, which occurred on August 14, 1868, and said that the house of Superintendent Harvey Beckwith, in which he and his wife were sleeping, slid part way into the cavern. In the next evening's paper the editor stated that he had received an indignant denial from the superintendent that he was a married man; therefore, he hastened to say that "the lady was not his wife."[19]

In 1875 the Chollar-Potosi joined the Hale & Norcross and the Savage in sinking the great Combination Shaft and began to levy heavy assessments, which continued until the shaft was closed on October 16, 1886. The assessments to date total $7,000,000. The Chollar and the Potosi were again segregated in 1875, and each given 700 feet of the Lode.

THE YELLOW JACKET

The Yellow Jacket was located on May 1, 1859, south of and adjoining the Little Gold Hill group soon after the placer miners had uncovered the top of the Old Red Ledge. There was no ore on the surface so the claim was laid over the large quartz croppings on the footwall side of the Lode.

The Princess and Union claims were located lower down the hill, adjoining the Yellow Jacket on the east. All went well until the discovery of a fine ore body in the Union tunnel, 90 feet beneath the surface.

---

and was awarded a degree by the University of Nevada. His one hundredth birthday was celebrated in 1939, which was the last. He was a stalwart State-of-Maine man; came to California at the age of seventeen; crossed the range to Nevada in 1859, where he built mills and superintendended mines for over thirty years; was elected Governor in 1891 and afterward served as Superintendent of the U. S. Mint at Carson City. He was a highly intelligent, honorable man, a Republican through thick and thin, and temperate in all things except smoking.

[19]Philip Lynch gave a humorous twist to his news items whenever possible.

The Yellow Jacket, it seems, then floated its claim down the hill, laid claim to the ore, and brought suit by its attorney, Wm. M. Stewart.

After the trial the Yellow Jacket began to sink a shaft 200 feet from its north end line, which entered a large body of rich ore at the depth of 180 feet. That deposit stood in the mass of quartz known as the Gold Hill bonanza, although it was not directly connected with any of the ore bodies. Later the ore was found to extend into the Confidence.

That mass of ore was 200 feet in length and 400 in height, and produced over $6,000,000. During the first year the ore milled $40 a ton, then fell to an average of $32.

"A sudden impoverishment" of the ore body above the 360-foot level in 1865 caused dismay. The report stated that while $3,000,000 had been produced, one half of that had been paid for milling and the company was $379,711 in debt. Meanwhile, $320,000 had been paid in dividends and $300,000 collected in assessments. All of the other producing mines appeared to be nearly worked out in 1865 and a panic followed. The Yellow Jacket, however, recovered almost immediately for the lack of ore on the 360-foot level was only temporary. In 1866 and 1867 a large tonnage of $32 ore was produced, but the ore body definitely ceased at the 475-foot level.

Sharon took control of the company in 1864 and it became the worst example of his management. Although the mine was in good ore from 1863 to 1872 and paid dividends every year it also levied assessments every year until 1871 in almost equal amounts. The dividends to that date totaled $2,184,000, and the assessments $1,558,000.

The annual reports of the other important mines go into great detail, while those of the Yellow Jacket are slight and give little information. The statements of assessments and dividends are quite undependable, as appears by a detailed list furnished by Geo. W. Hopkins, then secretary to James D. Hague in 1876.[20] The latter was unable to obtain more than partial estimates when he was writing his great book in 1869,[21] and seven years later was still dissatisfied.

Hopkins' list shows how simple it was to manipulate Yellow Jacket stock. Only the management knew when assessments

---

[20] The statement is at the Mackay School of Mines, which has the most complete collection of Comstock mine reports.

[21] Vol. 3 of *U. S. Exploration of the 40 Parallel.*

HOISTING WORKS OF LITTLE GOLD HILL MINES FROM THE IMPERIAL
TO THE CONFIDENCE, IN 1865

(See bottom opposite page)

were to be levied or dividends paid, and the price of the stock rose and fell accordingly. It was not Sharon's practice to pay assessments further than was absolutely necessary. "Sell when they are high and buy when they are low."

The company had a mill at Empire, but the greater part of the ore was reduced in "outside mills" (which meant Sharon's) at very high treatment charges. John D. Winters, superintendent from 1864 to 1870, said after a row with Sharon, that he had lost his self-respect doing dirty work for him. "To feed his mills I've mixed waste rock with Yellow Jacket ore until it would scarcely pay for crushing."[22]

In 1865 the Yellow Jacket began to sink its south shaft for the purpose of sharing in the ore body in the west-dipping vein discovered by the Crown Point in 1864, which extended through the Kentuck into the Yellow Jacket. The Crown Point also found the fine ore bodies lying along the hanging wall side of the Lode, which were more profitable to the Yellow Jacket than to the Crown Point or the Kentuck.

The mine was finished as a profitable enterprise in 1871 during which it paid $520,000 in dividends from the east ore bodies on the 1000-foot level. Thereafter its career was marked by one long record of assessments. Mackay and his partners took control of the mine in 1876 and sank the New Yellow Jacket vertical shaft 3,000 feet east of the croppings, which intersected the Lode at the depth of 2,800 feet. The shaft reached the depth of 3,060 feet, but found only barren quartz and floods of hot water. Two of the ablest men on the Comstock, I. E. James and Capt. Thomas G. Taylor, successively superintended that shaft, which cost about $2,000,000.

When it ceased pumping in 1882, the water in the Gold Hill mines gradually rose to the 1200 level where it has since remained. Control was turned back to Sharon in 1883, who, during the next ten years, extracted and milled about 350,000 tons of ore yielding an average of $11.90 a ton—all for the benefit of Sharon's mills and his railroad. When the ore did not pay the expense of mining and milling and transportation, assessments were levied.

---

[22]Miriam Michelson's *Wonderland of Silver and Gold*, p. 190 (1934).

HOISTING WORKS—These hoists, which indicate the small size of the claims, were built along the top of Little Gold Hill after it had been mined and leveled off. The Old Red Ledge, from which they obtained their good ore during the first four years, cropped along the top of the Hill and dipped westward, at 45 degrees, into the mountain. When that suddenly failed in 1863 the east ore bodies were found under the Hill and these hoists were built to sink small vertical shafts down to the ore.

Sharon ran the mine for years while holding only a fraction of the stock, so indifferent were Comstock stockholders.

During the past thirty years millions of tons of low grade, unprofitable ore have been mined from the surface of the Yellow Jacket and the Imperial, leaving a great hole.

The Yellow Jacket has produced over $20,000,000, including low-grade ore; its assessments total nearly $7,000,000.

The Original Gold Hill Bonanzas

In the fall of 1859, the top of Little Gold Hill was gashed with trenches and pits from which the owners were taking out disintegrated quartz and dirt to wash through their rockers and coarser quartz to reduce in arrastras and in the four-stamp mill that had been built on the Carson River. The newspaper reports of the earnings vary from $10 to $100 for each man. The poor placer miners who had laid locations along the top of the hill in January and February awoke a few months later to find themselves the owners of rich quartz mines after they dug down to the Old Red Ledge, which lay a few feet below the surface dirt. That remarkable ore body, which extended along the hill for nearly 500 feet and dipped west at an average angle of 45 degrees, was from 10 to 25 feet in width[23] and averaged $60 a ton mill returns, after a loss of about one third of the values in the tailings.

That little flat-topped hill, or mound as it was sometimes called, was the top of the Original Gold Hill Bonanza, which, in after years, proved to be 1,000 feet long, 500 feet wide, and reached an average depth of 500 feet, narrowing as it went down—although by far the larger part was unprofitable low-grade quartz and mineralized porphyry. A series of east ore bodies standing along the hanging wall side of the Lode was not discovered until nearly four years later.

Six months after the discovery of the quartz vein only three of the original locators retained their little claims: L. S. "Sandy" Bowers with a 20-foot claim, Joseph Plato with 10 feet, and Alec Henderson with 50 feet, and each of them acquired a substantial fortune. The others had been eager to sell to the first incoming Californians at comparatively small prices, just as practically all

---

[23]"The ledge is said to be fifteen feet wide. It is certainly eight to ten feet wide." (A. B. Paul in *San Francisco Bulletin*, June 12, 1860). Baron Von Richthofen, in his famous report of November 1865, stated the width of the ledge to be "from twelve to twenty-two feet."

of the other locators along the Lode had done.[24] The locators at Gold Hill received an average of $50 a foot for their claims. Such prices seem pitifully small in the light of later developments, but at the time they were fair enough. It was a remote and strange country and no one knew how long the rich ore would last.[25]

The longest of those little claims had but 50 feet along the Lode originally, and most of them were soon sold and subdivided into still smaller claims until there were twenty in all. One purchaser had only 8¾ feet in length, another 10 feet, several had 20 feet. In all, those twenty claims had less than 500 feet along the Lode, but fortunately each had a slice across that notable ore body. A number of those little claims were bought by others and consolidated. Three years later there were only eight little mines, excluding the Empire and the Imperial companies.

A primitive little settlement had sprung up just below the hill, similar in character to "Ophir Diggings," but even smaller. There were no frame buildings—only tents, rough cabins, and shanties of various kinds. Henry De Groot wrote that when he arrived on the Comstock in August of 1859,

> Mrs. Ellen Cowan was living at Gold Hill in a very rude and comfortless sort of an abode. She did the washing for the miners, a business that paid well at that day, and had gathered not a little gear prior to her marriage with Sandy.[26]

Sandy had but 10 feet of the Lode originally, and Ellen (as she is called in the record of mining claims) had bought an adjoining 10 feet from James Rogers, one of Sandy's colocators, for $100 a foot. She had visions of a rich ore body in that ground and Sandy soon found himself Mrs. Cowan's third husband. She was thirty-six, Sandy twenty-six.[27] No record of the marriage has been found.

---

[24]The names of the successive owners of the different claims are given in the *Nevada Directory* for 1862, p. 169, and in the *Nevada Directory* for 1863, p. 307. Nearly all of the claims had their own mills at some point in Gold Cañon or on the Carson River. The *Virginia Daily Union* of October 9, 1863, gives an authentic account of the discovery at Gold Hill and tells of the sales by the locators.

[25]Henry De Groot in *Mining and Scientific Press*, October 12, 1876.

[26]Id., November 25, 1876.

[27]Bancroft's History, Vol. 25, pp. 109, 110, quoting from the *Virginia City Union* of October 14, 1863, which gives an account of the discovery and of the subsequent fortunes, or rather misfortunes, of the locators. *Reno State Journal* of January 9, 1875 and January 5, 1878; Henry De Groot, in *Mining and Scientific Press* of November 25, 1876.

Bower's wealth has been a juicy morsel for sensational writers who knew little of the facts and cared less. His little claim, which was all he had, did not produce over $1,200,000, and probably not more than $1,000,000—of which not to exceed one half was profit. When Bowers died, on April 21, 1868, his entire estate, including mine, mill, Washoe mansion and furnishings, ranch and livestock, was valued by court appraisers at $88,998.[28]

Plato, an athlete, died from overexertion in 1863, leaving a rich widow. Henderson retained his interests until 1863, when he sold for a small fortune and returned to the East. Nearly all of the other locators were soon as poor as before. While Bowers and Plato had one of the richest portions of the Old Red Ledge, the east ore bodies in their claims were smaller and less continuous than those of their neighbors and were not profitable after 1865. When the Comstock was producing bountifully in April 1864, the Daily Union published a list of the important mines on the Lode, including the Little Gold Hill mines, giving a liberal estimated value of each per linear foot. The mines of Bowers and Plato were valued at $10,000 a foot; the Eclipse and two other claims on Gold Hill, which were in fact richer, were valued at $15,000. Sandy's 20 feet were hardly worth $200,000 at the time.

Sandy and Ellen were simple, kindly, unlettered folk, who were much preyed upon. They had visions of grandeur and built an elaborate mansion in Washoe Valley, which cost far less than $407,000 as claimed by several writers. Their trip to Europe lasted eleven months—not two or three years as the chroniclers have it.[29] After Sandy's death Mrs. Bowers told fortunes for a living, and was known locally as "The Washoe Seeress."[30] Old-time Comstockers would have smiled at "Eilley Orrum, Queen of the Comstock."

---

[28]The appraisement is on file in the County Clerk's Office at Virginia City. Plato's adjoining ten feet, which happened to be very rich, was worked through Bowers' shaft and reduced in his mill, and Bowers was generally credited with the yield of both.

[29]"Mrs. Ellison Bowers" so testified in the case of Hardenburg v. Bacon, 33 *California Reports*, p. 356 (Transcript of the evidence, 1865). Sandy called her "Allison" in his will. Mr. and Mrs. Bowers left San Francisco on the steamer Golden Gate on May 1, 1862, and returned the following April. D. O. Mills and family were also passengers.

[30]Dan DeQuille wrote in his sketch of "Snowshoe" Thompson that when Mrs. Cowan was living in Gold Cañon in 1858 she told Thompson to buy her a "peepstone" in Sacramento, which, she said, enabled her to find out all manner of things—rich mines, stolen treasure, and even to see the faces of the dead. *Overland Monthly* (2d Series) pp. 419–435. He failed to find such a stone.

Bowers was a Mason and well-respected. The Enterprise[31] said of him: "By his death the State has lost a good and useful citizen, and the working men of the country a true and sympathetic friend." His funeral was largely attended. Sandy's oft-quoted speech on the eve of departure for Europe may be apocryphal.[32]

The Old Red Ledge carried a higher percentage of gold than any other ore on the Comstock and provided the Little Gold Hill miners with a bountiful supply of ore during the first four years. Then, without warning, early in 1863, the vein was cut off as with a knife at an average depth of 275 feet, when it neared the east-dipping footwall of the Lode. It did not terminate against the footwall but on a nearly flat bed of clay. The mine owners were deeply discouraged for a short time until Superintendent R. N. "Bob" Graves of the Empire Company in March drove a cross-cut 300 feet to the east, on the 240-foot level of its McClellan shaft, and encountered several parallel vertical ore bodies standing along or near the east wall of the Lode.[33] The tops of these new irregular ore bodies were from 160 to 200 feet below the surface, but they extended through all of the little mines and yielded a large production for several years.[34] When those ore bodies terminated in most of the mines in 1865, at an average depth of 500 feet, the hoists were removed and the whole hillside along both sides of the Old Red Ledge was quarried for 200 feet in width and 200 feet in depth and sent to the mills. The average recovery was from $20 to $25 a ton, which yielded a fair profit.[35] In the Imperial, at the north end of the east bonanza, the good ore went down to 700 feet, and at the south end, in the Consolidated, to 600 feet.

Unlike the other Comstock mines, those little claims were owned and operated during the early years by one or more men

---

[31]April 23, 1868.

[32]Printed in Thompson & West *History of Nevada*, p. 622 (1881), and on pp. 160 and 161 of Shinn's *Story Of The Mine.*

[33]Report of Prof. Benjamin Silliman on the Empire, dated December 2, 1864.

[34]Although the east ore bodies averaged only $35 a ton and carried a much higher percentage of silver than the west vein (57%), the latter was assumed by some geologists to be a faulted top segment.

[35]Report of Prof. Benjamin Silliman on the Empire, December 2, 1864; *Gold Hill News*, October 22, 23, and November 11, 1866; also March 3, 1867; *U. S. Mineral Resources* for 1868, pp. 44–46; Id., 1869, p. 94; *U. S. Exploration of 40th Parallel*, Vol. 3, pp. 41, 51 (1870).

The site of that ore body is now marked by a great open cavern, over 1,000 feet long, 600 feet wide, and 400 feet deep, from which 2,000,000 tons of low grade ore were mined and milled during a long course of years, at no profit.

who preferred their rich independence to the doubtful advantages of incorporation and a treacherous stock market. Although their stopes and other mine workings constantly joined, the owners worked together harmoniously, often using each other's shafts. Those were almost the only mines along the Lode that were not involved in a welter of litigation, nor was any quantity of water encountered to interfere with operations. So small were the mines that their shaft houses stood side by side. Each had its own mill which made them quite independent. All in all those were such mines as prospectors dream of—little banks from which gold and silver can be drawn at will. But that easily won wealth was largely spent as it came and nearly all of the owners died poor.

In 1863, when the ore in the west ledge was nearly at an end, three of the small mines were incorporated into the Empire and three others into the Imperial, both of which paid dividends for five years on a moderate grade of ore mined from the east ore bodies.[36] The owners of several of the other eight little mines also prospered for several years, but by 1868 the last of them was done.[37]

In 1876 Sharon merged all of those mines into the Consolidated Imperial, which continued to sink the Imperial-Empire shaft and explore the Lode to the depth of 3,000 feet, finding only a very small body below the 1600-foot level. The Original Gold Hill Bonanza had yielded $25,000,000, and paid nearly $5,000,000 in dividends.[38]

---

[36]Those mines were managed in the interest of the stockholders by R. N. "Bob" Graves, superintendent of the Empire, and by P. S. Buckminster and his successor, J. D. Greentree, of the Imperial. R. K. Colcord, later governor of Nevada, was the construction engineer of the Imperial-Empire hoisting works and superintendent for several years.

[37]The fame of the Gold Hill section rested upon the original Gold Hill Bonanza during the first four years. The adjoining mines, which later became famous, did not begin to find rich ore until 1863.

[38]This is an estimate, reached after months of investigation. No reports were made by the private owners, and the historians give little information concerning those mines. The *San Francisco Chronicle* of August 5, 1877, printed a table purporting to give the production of each of the Comstock bonanzas, together with additional detailed information. While in some respects the table is obviously incorrect, it was prepared with care by some one familiar with the Lode, presumably by Squire P. Dewey. The Original Gold Hill bonanza is credited with a production of 1,037,413 tons, yielding an average of $25.39 a ton, or $26,340,762 in all, which is a close approximation.

CHAPTER XI

Three Years of Steady Progress, 1866, 1867, 1868—The Kentuck Brings Mackay's First Fortune.

The crash of the stock market in the fall of 1865, which culminated December 15, was fateful mainly to the few larger speculators who had been able to survive the decline of the two preceding years. Most of the smaller fry had been cleaned out long before. Among others who went to the wall was William M. Lent, one of the most prominent Comstock operators. "Uncle Billy," as he was called, recovered later and long remained an active figure in the mining world. While the rich ore in the Ophir and the Gould & Curry bonanzas had failed, and the Savage and the Yellow Jacket were temporarily in eclipse, the panic was an acute depression resulting from two years of financial strain.[1] The mines were worth far more than those bedrock prices.

Fortunately the darkness of December passed with the coming of the new year. During 1866 one mine after another found new and almost unexpected ore bodies, although none of the bonanza type. These new discoveries, together with the development of some additional ore in a number of the old mines, gave the Comstock a fair measure of prosperity for three years. Men's expectations had moderated; they had learned that bonanzas, however rich, were exhausted within a few years.[2] The stock market improved and remained fairly steady for three years, then slumped again during the gloomy year of 1869 and hit bottom in the still darker year that followed. The number of men employed in the mines is often overestimated. There were only 1,450 miners in February 1866, producing 1,453 tons of ore.[3] The average number of men employed underground during the years 1866, 1867, 1868 was about 1,500, and the average daily production of ore from the few producing mines was 1,500 tons.

The hoisting and pumping appliances in the second-line vertical

---

[1]Mining throughout the rest of Nevada was in a state of paralysis. The total production for 1866 outside of the Comstock did not exceed $500,000. *U. S. Mineral Resources* for 1866, p. 124.

[2]The exhaustion of the rich ores during 1865 caused production to drop to $12,000,000 in 1866, a decrease of $4,000,000. During 1867 it rose to $13,738,618, and in 1868 stood at $12,418,000. Known ore bodies were failing fast in 1869 and production fell to $7,500,000, with no prospect of an increase.

[3]*Gold Hill News*, February 17, 1866. Lynch visited all of the mines and tabulated the number of miners in each.

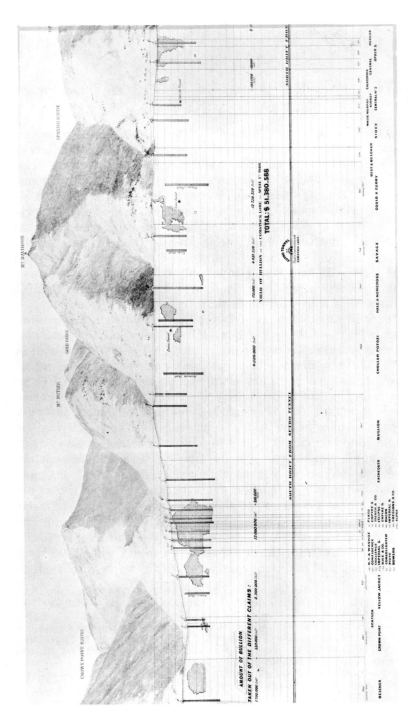

Depths of shafts and indicated bonanzas at the time of the *proposed* Sutro Tunnel, 1866, as visualized by Adolph Sutro.

shafts were looked upon as marvels,[4] but they were toys compared with the machinery installed at the third-line vertical shafts ten to twelve years later. The second-line shafts were usually from $4\frac{1}{2}$ to 6 feet wide and from 20 to 24 feet long, inside the timbers, and divided into four compartments; two were used for hoisting, one for sinking and one for pumping: Machine drills and giant powder were yet to come. The ventilation of the mines continued to be a major problem. Root blowers, driven by small engines, were in general use for forcing air down the shafts. The newspaper heading "Terrible Mining Accident" appears all too often.

Most of the mines had the water under control during the late '60s, owing to the installation of larger pumps and to the fact that only a few of the shafts exceeded 600 feet in depth. The unfortunate exceptions were the Gould & Curry and the Ophir where the battle was unceasing. Both mines were out of ore and bending all efforts upon sinking their main shafts in the hope of finding other bonanzas. The managers were almost discouraged to find that as the shafts gained in depth the volume of water increased as fast as additional pumps could be installed.

The accompanying vertical section of the Comstock Lode in 1866, which was published by Sutro to promote the sale of Tunnel stock, tells more than words how few ore bodies had been found up to that time, and nearly all of them were practically exhausted. Less than one-five-hundredth part of the Lode was payable ore, although none of the quartz and vein matter is quite barren. The section may be misleading, as nothing whatever was then known below the bottoms of the shafts which are represented by heavy black lines.

While the Comstock was no longer as picturesque and lively as during the boom years of 1863 and 1864, life was never dull. The activity on the streets was continuous and the interest and the cheerfulness of the people never flagged, although living conditions were hard for the majority, particularly during the winter. Their homes were small frame houses in which many children were packed like sardines. Wood was the only fuel; kerosene and candles provided the lights except in the business district and nearby large homes, which were served with gas. The only modern improvement for the housewives was the sewing machine—a

---

[4] Methods of timbering and the details of mining and pumping machinery in 1869 are explained and illustrated by James D. Hague in Vol. 3, pp. 102–146 of *U. S. Exploration of the 40th Parallel* (1870). The Cornish pumping plant that he illustrates on page 146 grew into a mammoth.

luxury not enjoyed by many. Water was piped to the central part of the town, but was delivered in barrels to many little houses on the outskirts. Nevertheless those people found more zest in living than modern city dwellers with all their conveniences. The deeper satisfactions of life do not lie in comforts.

The traffic down that steep narrow road through Gold Hill to the mills in Gold Cañon and on the Carson River can be imagined from the fact that in February 1867 the Gold Hill News described a jam of 65 loaded ore wagons, drawn by 325 horses, in the lower main street of Gold Hill, all caused by an accident to the first wagon and the coming-on of the others. The road was a river of mud and the wagons slid on despite the brakes. "Slippery Gulch" itself.

The winters of 1866–1867 and 1867–1868 were the heaviest ever known in the Washoe region. Although hundreds of men were shoveling snow on the highways over the Sierras the stages did not get through for ten days. Mills and mines on the Comstock were at a standstill after a snowfall of four feet in February which blocked the hauling of ore. Many men were thrown out of work and their families suffered for want of food and firewood. The more fortunate organized a "Relief Committee."

The number of local and outside newspapers read daily by those people brought the comment from Lord: "This passion for news has never been more general and intense than in the Comstock district." He tells also of their interest in periodicals and other good literature in 1867.

The Central Pacific Railroad was completed to Cisco, near the crest of the Sierras, in November 1866. The company had built a good highway ahead of the railroad over the Donner Lake pass in the fall of 1863, and had been steadily diverting the freight and passenger traffic from the Placerville and other mountain roads after the railroad reached Colfax in 1865. In June of 1867 the last stage and fast freight wagons were withdrawn from the Placerville road, which was no longer maintained as a great highway.

"A train of six camels came in yesterday with loads of charcoal from a ranch near Fort Churchill, some accompanied by their young," reported the Gold Hill News of May 14, 1867. These were the survivors of the small herds brought from California in

1861 to carry salt from the deserts to the mills and were owned by a Frenchman at Fort Churchill.[5]

THE KENTUCK BRINGS MACKAY'S FIRST FORTUNE

Mackay came up the hard way—by working with his hands for interests in mines. But none of them looked promising in 1863. The Union was still a prospect, although well located; the Caledonia Tunnel was a failure, and the Milton, of which he had become superintendent, conflicted with the Chollar. The Milton claimed to have the earlier location and actively developed its mine for several unfruitful years, after which it was absorbed by the Chollar. Mackay had built up a great asset in those four early years; he had begun to impress men with his ability, his uprightness, and his steady judgment,[6] which led to his association with the Bullion.

The Bullion, then claiming 1,424 feet of the Lode covering "The Divide" and lying between two great ore bodies—the Chollar-Potosi and the Gold Hill Bonanza—was incorporated in 1863 by J. M. Walker and fifty-five coowners,[7] who were preparing to sink a deep shaft and explore the Lode in search of the great ore bodies that all believed to exist there. Walker, one of the principal owners and the superintendent, invited Mackay to join the enterprise, who welcomed the opportunity. He disposed of his other interests, no doubt to advantage in that wild market.

Walker had offered a partnership embracing all of their interests and they operated together for four years—probably by verbal agreements, as the firm of Mackay, Fair, Flood, and O'Brien did afterward. Mackay and Walker did not own the control of the Bullion but retained the management by consent of the stockholders, Walker having the title of superintendent and Mackay without official authority except as one of the five trustees, but it is probable that he was the mainspring of the organization. Walker was an agreeable gentleman from the State of Virginia, but not a forceful character nor a man of much practical ability, as his after life showed.

---

[5] An informative monograph on "Camels in Western America," by A. A. Gray, Francis P. Farquhar and William S. Lewis, appeared in the *Quarterly of the California Historical Society* for December 1930.

[6] D. O. Mills in *New York Herald*, July 21, 1902; *San Francisco Chronicle*, July 21, 1902.

[7] The number of owners of interests in the Bullion illustrates the extent to which men traded in mines in the early days.

The Bullion was equipped with the most modern machinery, and the shaft was sunk more rapidly than any other on the Comstock. As sinking progressed the Lode was thoroughly explored for a width of 200 feet on level after level. Everywhere the workings disclosed masses of white low-grade quartz in which ore bodies were commonly found, but none of it was rich enough to be called ore. At times the indications were most promising and hopes were high. A few feet more, they thought, or perhaps on the next level. It was a gallant fight. Assessments of $10 a share were levied like clockwork every ninety days upon the 2,500 shares in the mine. The stockholders, in the main, were local people. As some dropped out others took their places. This was almost the only important mine of the period, with the exception of those at Little Gold Hill, that was not controlled in San Francisco.

When the Kentuck became profitable in the fall of 1867, Walker sold his stock to Sharon and disposed of his other Comstock interests, and left for an extended stay in Europe. Mackay was then elected superintendent.

The Bullion, however, defied all efforts. On August 29, 1869, the editor of the Gold Hill News[8] regretfully announced that the Bullion shaft was 1,400 feet deep, and dry, with no ore and no indications of ore.[9] The Enterprise reported that the bottom of the shaft was "as dry as a lime-kiln and as hot as an oven."[10] That was the only deep shaft on the Lode that did not encounter great volumes of water, which probably accounts for the lack of ore. It would seem that the Lode where it extended through the Bullion, and for several hundred feet beyond each end was too "tight" to give entrance to the hot waters and gases which deposited the rich minerals in other portions of the vein after it had been filled with masses of low-grade quartz.

---

[8]The *Gold Hill News* printed many articles from 1863 to 1869 telling of progress at the Bullion.

[9]Late in 1867 the other deep shafts were: Gould & Curry, 850 feet; Savage, 670; Hale & Norcross, 600; Chollar-Potosi, 830; Imperial-Empire, 920; (*U. S. Mineral Resources* for 1867, p. 343). All of these were vertical second-line shafts, which were turned into inclines when they reached the footwall of the Lode.

[10]*Daily Territorial Enterprise*, August 7, 1868.

It is interesting to note that the rock was hot despite the absence of hot water. In later years, after the shaft reached a depth of 2,000 feet, floods of hot water greatly impeded operations.

The Ward shaft, the third-line shaft of the Bullion and the Julia, was stopped at 2,725 feet by running ground and intense heat. It did not reach the Lode.

A million dollars had been spent on the mine without producing a ton of ore, and Mackay and his associates were content to quit. The Bullion then passed into the control of a new crowd that made a football of the stock and collected $2,000,000 in assessments during the succeeding fifteen years, without finding ore. It happened that the mine lay in the central portion of a section of the Lode, 1,800 feet in length, which proved barren to the deepest level. The record of the Bullion illustrates the enormous sums spent on unprofitable mines. From 1863 onward, the various managements that controlled it levied and collected over $5,000,000 in assessments.

The depression of 1865 was opportune for Mackay and his partner, J. M. Walker—it made the Kentuck available for their limited capital. That little mine, with only 94 feet of the Lode, had never produced a ton of ore, yet at the time it was the surest venture ever known on the Comstock. Always before it had been a gamble whether an ore body would be found in an undeveloped mine, but in this instance the Crown Point had recently found its first ore body in the west-dipping vein on the 230-foot level, almost at the Kentuck's south line. It was a practical certainty that the ore would extend into the Kentuck and perhaps through it, as the Yellow Jacket, which adjoined on the north, had recently encountered an ore body with the same course and the same westerly dip.

No stock was offered to the public and there was no blowing of horns. A small shaft was started immediately, with a horse-whim for a hoist. As befitted a small group of owners the work was done quietly and economically.

Fortune smiled and turned her wheel this time. On January 3, 1866, the Gold Hill News reported that on January 1 the Kentuck struck ten feet of ore in the shaft at the depth of 275 feet. Now it was assured that the ore extended from the northerly portion of the Crown Point through the Kentuck and into the Yellow Jacket.

This was another of those peculiar ore bodies in the west-dipping vein. Where the ore was first found in the Crown Point shaft it was 15 feet thick. Above that point it narrowed rapidly until it terminated in a thin blade of ore 90 feet below the surface. Below the 230-foot level the ore increased in width until it was from 40 to 65 feet wide at the 400 level, where it bottomed on a nearly horizontal bed of clay as it neared the easterly dipping footwall of the Comstock Lode.[11]　The length of that ore body was 500 feet and the three mines extracted from it about $3,000,000.

The Crown Point was the first to find that the west vein had been cut off. Superintendent C. C. Batterman[12] then drove a crosscut eastward on the 500-foot level through the great "horse" that lay in that portion of the Lode to the east wall, where it encountered the top of a large and rich easterly-dipping ore body that pitched through the Kentuck into the Yellow Jacket, and provided dividends for those mines for three years.[13]　The Kentuck now became one of the most profitable mines on the Lode. Always well managed, it soon began to operate through the shafts of its neighbors, and continued to do so ever afterward. For so small a mine the output of ore was extraordinary. From January 1, 1866, to November 1, 1869, the gross production was $3,641,062, from which $1,142,000 was paid in dividends;[14]　no assessments being collected in the meantime except one of $40,000 following the Yellow Jacket fire of April 7, 1869, which cost the lives of thirty-seven men and wrecked the adjoining stopes of the Crown Point, the Kentuck, and the Yellow Jacket.

It was wormwood to Sharon to see a prosperous mine and mill on the Comstock that he did not control, so he maneuvered to acquire the Kentuck. The Chollar-Potosi, the Crown Point, and the Yellow Jacket were already in his hands. Walker wanted to retire and go to Europe so his partnership with Mackay was terminated, Walker's share being $600,000.[15]

---

[11]All of the ore bodies found in the west-dipping vein in the Little Gold Hill Mines had been cut off in the same way.

[12]C. C. Batterman became one of the most prominent men in the mining, social, and political life of Nevada. The title of "General" heightened the natural dignity of his small rotund figure.

[13]The geologists divided in opinion whether the west-dipping ore body was a faulted segment of the east ore body. Rossiter W. Raymond advanced the theory in *U. S. Mineral Resources* for 1868, p. 46; Id., 1869, p. 94; Clarence King disagreed, in *U. S. Exploration of the 40th Parallel*, pp. 49, 50 (1870); Becker states it as a fault in his *Geology of the Comstock Lode*, p. 277 (1883). The east ore bodies terminated at 800 feet in the Crown Point, 900 in the Kentuck, and at 1,000 feet in the Yellow Jacket.

[14]Mackay's first success came when he obtained control of the Kentuck mine, which became a dividend producer under his management and out of which he made a good deal of money. Marye's *From '49 to '83 in California and Nevada*, p. 109 (1923).

[15]James E. Walsh. Walsh was manager of the Flood estate and connected with the Flood family and with Mackay for a lifetime. He held a power of attorney from Mackay for ten years prior to the latter's death, to handle and convey real estate in California. The passing of Walsh, some years ago, was a distinct loss to San Francisco.

CHAPTER XII

THE SUTRO TUNNEL

Adolph Heinrich Joseph Sutro was one of the most remarkable men that rose to power on the Comstock. For all-round ability he had no superior. Although speaking broken English to the day of his death, he was a fluent and convincing speaker and writer, and so quick and shifty in argument as to discomfit clever lawyers. He was a dynamo of energy, sleepless in effort, and a first-class fighting man. The pertinacity of the Jewish race was exemplified in him. Twice he prevailed upon the Committee on Mines and Mining of the House of Representatives to recommend a bill providing for a Federal loan to construct the tunnel, but each time the bill failed to become a law.

Although hated, almost despised, by many on the Comstock he peddled his stock in the East and in Europe, fought some glorious battles against the combined mining, financial, and political powers of the West, and finished his tunnel in 1878, after thirteen years of effort. It was too late, the last great ore body, the Con. Virginia, was practically exhausted, a number of the mines had already sunk their shafts far below the tunnel level without finding ore, and the tunnel appeared to be, as it was, a financial failure.[1] No one knew that better than Sutro. The only source of income was the royalty on ore, and there was little of it left. He quarreled with his directors over the proposed compromise with the mines to reduce the royalty from $2 to $1 a ton, resigned from the company, and began to sell his large block of stock, whose price gradually fell from $6.50 to 6¢ a share. Meantime he had retired to San Francisco a millionaire. This is admitted in the elaborate sketch of his life in The Dictionary of American Biography, which says that after the completion of the tunnel, "The project proved immediately and immensely profitable" (so much for biography), and that, "In 1879 Sutro sold his interest

---

[1]The story of the tunnel is told at length in Thompson & West *History of Nevada*, pp. 505–511 (1881) ; Lord's *Comstock Mining and Miners*, pp. 233–243, 333–344 (1883) ; Shinn's *Story of the Mine*, pp. 194–208 (1896), and in Vol. 25 of *Bancroft's History*, pp. 141–149 (1890). Judge Goodwin gives a sketch of Sutro in *As I Remember Them*, pp. 240–244. All of those writers pay high tribute to Sutro's accomplishments, although criticizing his methods.

Joseph Aron, to whom Sutro paid a great tribute before Congress, *Sutro Tunnel*, p. 943 (1872), issued a large pamphlet 20 years later bitterly criticizing Sutro's treatment of his stockholders. *History of a Great Work and an Honest Miner*. The Bancroft Library has a copy.

and returned to San Francisco where he invested his tunnel profits in real estate."

Early in 1864 the desirability of a deep tunnel to drain and explore the entire Lode had become the subject of a good deal of discussion in the newspapers and otherwise. The Gold Hill News of March 14, 1864, had an editorial on the necessity for a tunnel five or six miles in length, with lateral tunnels driven north and south from the main tunnel to drain all of the mines. Sutro supported the plan, for he was a shrewd, farseeing man; his Dayton mill had burned, and he was in need of an enterprise. But that he was the first to suggest the idea to those active-minded people is most improbable. He quickly became its champion and carried the enterprise to completion, which is glory enough.

Sutro plunged into the enterprise with his accustomed energy and on February 4, 1865, the Nevada Legislature passed an Act granting to "A. Sutro and his associates" an exclusive franchise to construct and operate the tunnel for a period of fifty years. Senator Stewart became the first president and the enterprise had the enthusiastic support of all the mines, the Bank of California, and many leading citizens of Nevada and California.[2] In April 1866, twenty-three of the leading mining companies entered into an agreement to pay the Tunnel Company a royalty of $2 on every ton of ore extracted by each mine after the tunnel began to drain it. The mines were given the right also to use the tunnel for the transportation of men, supplies, etc., at specified rates.

The financial set-up of the tunnel was wrong from the first. If the mining companies wanted to prospect and drain the Lode at great depth they should have formed a corporation and taken stock and financed it by assessments. In that event it would have been finished in time to be of use.

But when Sutro took up the enterprise he dominated it. Clever man that he was, he proposed to raise the necessary $3,000,000 in the East. The Comstock was in the midst of a depression. The upper ore bodies had failed at an average depth of 500 feet and

---

[2]Lord's *Comstock Mining and Miners*, p. 233 (1883). A long list of the most influential men in Nevada and California endorsing the project is printed on pages 183–198 of Sutro's first book *Sutro Tunnel, Nevada* (1868). The list includes mine superintendents, State officers, judges, and business men of Nevada; also the names of the mines favoring it, together with the names of the presidents, trustees and secretaries, nearly all of whom resided in San Francisco.

no one knew what might be found below. A tunnel to tap the Lode 1,200 feet deeper was a gamble. Let the easterners take the risk. If ore was found the companies could well afford to pay a royalty of $2 a ton when their mines were drained. It seems that Sutro kept control of the stock from first to last.

Senator Stewart induced Congress to pass the Mining Act of July 25, 1866, the first Federal law to provide for a location and patenting of mining claims on public lands. On the same day, the Sutro Tunnel Act was passed, which granted the company the right to construct the tunnel, giving it the exclusive ownership of all lodes and ledges discovered along its course for 2,000 feet on each side for a length of seven miles,[3] excepting, of course, the Comstock Lode, which the tunnel eventually intersected 20,589 feet from its mouth and 1,750 feet vertically beneath the surface.

Sutro thought it would be a simple matter to sell that stock in the East, but the New York capitalists were shy. They were well aware of the depression and suggested that he first obtain subscriptions to the amount of $400,000 or $500,000 in the West. Sutro returned in 1867 and secured subscriptions amounting to $600,000 from eleven of the mining companies, which also granted an extension of one year's time in which to commence operations. The other twelve companies, which had signed the agreement a year earlier, part of which were controlled by Sharon, declined to renew their contracts and to subscribe for stock, on the excuse that Sutro had not complied with his agreement. The value of the ore mined had fallen to $30 a ton, on which a royalty of $2 a ton was thought excessive.

Meantime, by 1867, a good deal of opposition to Sutro had developed. A number of prominent business men of San Francisco who were heavily interested in the Comstock had known Sutro when he was engaged in business there and were vigorous in their criticism of him and his plan.[4]

Sutro's vanity, aggressiveness, and "insufferable egotism" made him personally offensive to the leading Comstockers. Then he drove the whole region into opposition by proclaiming that the tunnel, and himself, of course, would dominate the Lode; all of the mining, he said, would be done through the tunnel, the mills would be located at its mouth, and Virginia City would become

---

[3]The original plan was to extend the tunnel westward under Mt. Davidson far beyond the Comstock Lode.

[4]This is revealed in Sutro's book, *Sutro Tunnel, Nevada*, p. 93 (1868).

almost an abandoned town. "The owls would roost in it," he said.

Sharon was the leader of the opposition on the Comstock, and with the active support of the powerful Bank of California thwarted all of Sutro's efforts to obtain large subscriptions for stock in the West and in the East as well. He did sell more or less stock on which to live and carry on his fight. Sutro was unceasing in his attacks on Sharon, on whose "kwooked" methods he never ceased to orate.

Baffled on every hand, Sutro returned in December 1867, resolved to appeal to Congress for a loan to complete the tunnel on the ground that it was a public necessity; the tunnel must be completed in order to save the Comstock. He painted a dark picture of conditions at the mines. The richest Lode on the earth was in the grasp of a ring of crooked manipulators who were plundering it for their own advantage. Soon the ore in sight would be worked out, floods of water would prevent deeper exploration, and the flourishing Comstock would be but a memory, with its untold wealth unattainable. He protested that he was a lone man, without means, fighting the battle of the people and desiring only to preserve a great heritage.

Sutro's arguments fell upon willing ears. The House of Representatives was resentful that a solemn Act of Congress should be flouted, and deeply moved by Sutro's representations of villainy. He haunted Congress for three years, cultivated the members of the House assiduously, and gained warm supporters. Two bills were presented granting subsidies, which failed to pass.

The disastrous fire in the Yellow Jacket, which took the lives of thirty-seven miners in April 1869 gave Sutro his opportunity. He was a stirring advocate and made the most of that calamity; proclaiming that the men would not have died if the tunnel had been completed, as it would have ventilated the mines and afforded an avenue of escape. He delivered lectures illustrated with lantern slides, showing the burning mines and dying miners, and, in reverse, the great Sutro tunnel by means of which all of the lives would have been saved. So aroused were the miners that the Miners' Union subscribed $50,000, with which the tunnel was started in October, notwithstanding vicious newspaper attacks.[5]

---

[5]The *Gold Hill News* editorially denounced Sutro personally as a questionable character, and his enterprise as "one of the most infamous and bare-faced swindles ever put forth in Nevada." (*Gold Hill News*, September 27, 1869.) Specimen newspaper attacks upon Sutro may be found in Vol. 2 of Bancroft's *Nevada Clippings*, pp. 520–537, at the Bancroft Library.

Sutro was nearly discouraged in the late '60s when he was thwarted on every hand. The tunnel was all but closed down toward the end of 1870; only 1,750 feet, of small dimensions, had been dug, and that through easy ground. The Crown Point discovery at the end of that year saved the tunnel as well as the Comstock; now there was no denying that great ore bodies were to be found below the 1000-foot level. Sutro was rejuvenated. Congress had passed an Act providing for a commission to examine into the project with a view to a Government loan, which gave him another strong talking point.

The first substantial subscriptions for stock was obtained in England, in September 1871. With the assistance of influential agents, 200,000 shares of stock were sold for $650,000 to the banking firm of McCalmont Brothers & Co. of London; three fourths of which was taken by McCalmont Brothers and the remainder by the Seligmans. This sum was followed by another subscription of $800,000 from European sources soon afterward.

Sutro now surrendered hope of securing a Federal loan and returned to devote all of his energies to the tunnel. It must be completed before the ore was exhausted. He adopted every mechanical device to hasten the work, and his untiring spirit animated the miners who performed prodigies of labor while combating excessive heat, foul air, floods of water, and unstable treacherous ground. Four shafts were started at intervals of 4,000 feet along the line of the tunnel to afford ventilation and to enable sections of the tunnel to be driven in advance of the main heading. Two of these were abandoned because the pumps could not handle the water. So difficult was the work that it took seven years to reach the Lode.

In 1873, after the Crown Point-Belcher bonanza was yielding millions and the Con. Virginia bonanza had been discovered, Sutro went to London and persuaded McCalmont Brothers & Co. to agree to float a bond issue of $7,500,000 to complete the tunnel.

The elaborate prospectus that Sutro issued in support of the bonds entitled "The Sutro Tunnel and Railway" was a masterpiece of its kind. No other ever equaled it, and all of its claims proved false. The mines, he said, contained $500,000,000 in lowgrade ores, which could be worked at a profit through the tunnel. (This is $100,000,000 more than the Comstock produced in all its history.) The great and rich veins to be discovered in the tunnel would amaze the world. All in all, the annual revenues for many years to come he tabulated at $22,076,125. An inviting prospect truly, but during the first five years after the tunnel was completed they averaged $44,000 a year, rising to $100,000 or more a year during the low-grade period and no rich veins were encountered in the tunnel.

Sutro was now to suffer another deep disappointment. Those Scotch bankers balked at selling bonds supported by the incredible statements in that prospectus and withdrew the issue. The McCalmonts were stockholders by reason of their purchase of shares in 1871; and, in 1877, when money was needed to complete the tunnel, they began to lend the Tunnel Company various sums on mortgages at 12 percent per annum.[7]

Sutro became a candidate for the United States Senate from Nevada in every election for that office from 1872 to 1880 and did not receive a vote before any of the legislatures, although the money bags of the various candidates were untied.[8]   Tom Fitch was recalled by Sharon to make speeches in his behalf in the campaign of 1874.   Fitch abused and ridiculed Sutro as only he could.   His closing address, delivered in the Opera House on October 23, 1874, was a great oration, interlarded with poetry, historical references, and flights into the blue.   It fills six closely printed columns of the Enterprise of October 24, under the heading, *"Sutro Annihilated—His Contracts All Violated—His Robberies Exposed."*

Sharon and Fitch were vulnerable,[9] and Sutro filled them full of poisoned arrows in an address delivered in the Opera House a few nights later.   Again he brought into play his famous lantern slides and lampoons.   His "lecture," illustrated, fills four pages of a supplement to his own newspaper "The Daily Independent" of October 31, 1874.[10]

The tunnel made the first connection with the workings of the Savage on July 8, 1878, and reached the Savage shaft on September 1.   The total length was 20,498 feet.   For the greater part of the distance the tunnel, inside of timbering, was 7 to 7½ feet in height, 8 feet wide across the top, and 9 to 9½ feet across the bottom.   The actual cost of construction up to that time was

---

[7]The first mortgage was filed February 7, 1877.

[8]Thompson & West *History of Nevada*, p. 92.

[9]Fitch was irrepressible. When a Republican convention failed to nominate him for Congress, he rose and said: "Gentlemen: from the bottom of my heart I can now sympathize with Lazarus—I too have been licked by dogs."

[10]The Bancroft Library has a copy; the Huntington Library at San Marino another.

$3,500,000, according to Theodore Sutro.[11]   Lord says the report of Secretary Ames shows the cost to be $2,096,566 up to 1881, not counting the expense of management.[12]

Then followed months of wrangling with the mining companies, which refused to pay royalty on the ground that the contract had been abrogated years before.   The storm broke when the Hale & Norcross pump rod broke, and that mine and the Savage began to pump water into the tunnel, driving out Sutro's men.   He retaliated by attempting to send the water back again and was stopped by a court injunction.   Then he threatened to bulkhead the tunnel.   Soon afterward, on March 1, 1879, a new agreement was reached whereby the mines were to pay $1 a ton royalty on ore mined not exceeding the value of $40 a ton and $2 on ore exceeding that value.[13]   It was at this time that Sutro resigned.   Until a mine was drained it was not required to pay a royalty.   Lateral tunnels to drain the various mines along the Lode were then started, and, in the course of several years, the north lateral, 4,403 feet in length, reached the Union shaft.[14] The south lateral was extended 8,423 feet to the Alta shaft. Lord says that after the completion of the laterals the quantity of water flowing through the tunnel varied from 3,500,000 to 4,000,000 gallons daily.[15]   The great Cornish pumps at the various shafts were raising more water than that before the laterals reached them and could have handled the water from the deep levels without the aid of the tunnel, although at greater expense.

If the tunnel had been completed in the early '70s it would have saved the mining companies millions of dollars in pumping expense, but, by the time the lateral tunnels reached the various shafts, a number of them had attained their greatest depth, especially the Union shaft and the Gold Hill mines.   Several of the shafts were 1,500 feet deeper than the tunnel when it first reached the Lode.   No ore was discovered in or by means of the tunnel, nor did the air serve as a good means of ventilation after passing through that hot wet tunnel.

During the past fifty years the tunnel has automatically drained the Gold Hill mines, where hot water of 150 degrees rose to the

[11]"The Sutro Tunnel Company and the Sutro Tunnel" (1887).

[12]*Comstock Mining and Miners*, p. 342 (1883).

[13]Id., p. 340.

[14]The mining companies advanced $70 a foot for driving the lateral tunnels, the advance to be credited on future royalties.   The north lateral of the tunnel was vertically 1,650 feet below the surface at the C. & C. shaft; the south lateral 1,435 feet below the collar of the New Yellow Jacket shaft.

[15]A large covered drain-box lay underneath the tracks.

SUTRO TUNNEL   SUTRO TUNNEL AND MULE TRAIN
ADOLPH SUTRO, FOUNDER OF THE SUTRO TUNNEL
TOWN OF SUTRO   SUTRO MACHINE SHOP AND MANSION

tunnel level. In the Virginia City mines the water has stood about 100 feet below the tunnel since 1884, with the exception of the brief period after 1900 when the North End mines were pumped out to the 2500 level.

The tunnel was a burden for five or six years after its completion, as the stock was nonassessable. When the royalties became substantial during the low-grade period the McCalmonts brought suit to foreclose their mortgages, upon which there was due, in principal and interest, the sum of $2,023,833. No part of the cost of the tunnel was ever repaid, with the exception of about $1,000,000 which McCalmont Brothers & Co. are said to have received on their mortgages, on a compromise, when the property of the Sutro Tunnel Company was sold on foreclosure in 1889. A large number of the old stockholders of the company had financed the arrangement and a new company, called the Comstock Tunnel Company, took over the property.[16]

Sutro's immense energy did not flag after his removal to San Francisco. He bought and improved a large area west of the city proper and forwarded the growth and improvements of the western area of San Francisco. To the very last Sutro craved distinction. During the hard times of 1893 and 1894 he acquired political prominence by attacking the corporations, and, in the latter year, was elected Mayor of the city "on the Populist Ticket," so his biography says. He died in 1898.

---

[16]Theodore Sutro of New York, said to be a brother of Adolph, was active in arranging for that reorganization. He published a book, "The Sutro Tunnel Company and the Sutro Tunnel" (1887), in which the history of the Tunnel is recited, without mention of Sutro's name. Theodore said he had lost money on the original enterprise.

CHAPTER XIII

The Hale & Norcross Venture of Mackay, Fair, Flood, and
O'Brien in 1869—They Join the Water Company Which
Brings Water from the Sierras.

While the Hale & Norcross lay between two rich and prosperous mines—the Savage and the Chollar-Potosi—the Lode at
that point was split by a wide porphyry "horse," and the large
bodies of quartz on the footwall side, which were first explored,
would not pay to mill in the early years. Assessment followed
assessment until December 1865 when a rich ore body was discovered on the 600-foot level in a crosscut to the east side of the
"horse" by Superintendent C. C. Thomas. Thereafter, during
1866 and 1867, the mine produced $2,200,000 and paid $790,000
in dividends. James G. Fair, who had been superintendent of
the Ophir for a brief period, apparently became foreman or perhaps assistant superintendent of the Hale & Norcross early in
1867, but, for some reason not disclosed, he lost the position on
November 28, 1867.[1] Fair was deeply incensed at his discharge,
and left the Comstock the following spring for Idaho, where he
remained until the latter part of 1868.

Sharon coveted the Hale & Norcross ore and set to work to
take control of the mine from C. L. Low and his friends, who
had seen it through hard times. A bitter and spectacular contest ensued that rocked the San Francisco Stock Exchange for
months. The 800 shares in the company rose to fabulous prices:
from $300 a share on January 8, 1868, to $1,475 on the 10th, to
$2,200 on the 11th, to $2,900 on the 13th, to $4,100 February 11,
and to $7,100 on February 15. This unprecedented increase was
due to the fact that the contending factions grimly held their
stock and that a number of brokers who had sold shares "short"
before the rise could not obtain stock with which to "fill." On
February 12 two shares sold for $10,000 each, and during one
week, at the informal sessions of the Exchange, $100,000 was

---

[1]*Gold Hill News*, November 30, 1867. Fair did not come to the Comstock
until 1865. He was already an experienced quartz miner. The Ophir report
of 1867 criticized him for taking away the plans of the new shaft. He claimed
to have been superintendent of the Hale & Norcross during 1867 and the early
part of 1868, but the annual reports for those years show C. C. Thomas to be
superintendent.

publicly offered each day for ten shares. The "shorts" were fearfully punished.[2]

At the annual meeting of stockholders held March 10, 1868, Sharon elected his board of directors, but won a barren victory. His mills got little ore from the mine that year. Three assessments amounting to $201,960 were levied, and the stock, which had been increased from 800 to 8,000 shares, fell below $50 before the end of the year.

Fair now returned from Idaho and convinced Mackay and Flood that the property could be made to pay. It had been extravagantly managed and the yearly reports showed a large volume of ore of moderate grade in the old upper workings; besides, the lower levels were by no means exhausted. Mackay's familiarity with the mine gave assurance that the investment would not be lost in any event, so he and Fair and James C. Flood and William S. O'Brien entered into a verbal agreement to make an effort to obtain control of the stock—Mackay taking a ⅜ths interest in the association, Flood and O'Brien ⅜ths, and Fair ⅔ths. To carry their share of the deal, Flood and O'Brien were forced to borrow something in excess of $50,000 from Edward Barron of San Francisco, and they, together with Mackay, helped Fair to carry his share.[3] Mackay took the larger interest because he was already wealthy, measured by those times. That verbal agreement of partnership, and that proportion of interests, continued for years, involving transactions running into hundreds of millions.

Flood, who had been dealing in mining stocks for several years with some success,[4] quietly set about the purchase of stock. As the mine had been levying assessments Sharon did not have a majority, and before he and his friends realized what had happened the new firm had control of the stock.[5] They nearly missed it, however, for Flood, innocently and as a measure of precaution, had put the purchased stock certificates in his safe without transferring them on the books. Not until the day before the

---

[2]King's *History of San Francisco Stock Exchange*, p. 39. To enable the shorts to make settlements the stock was not called on the Board for two months.

[3]James E. Walsh.

[4]Flood and O'Brien never lived on the Comstock. They ran the Auction Lunch saloon on Washington Street in San Francisco from 1852 until 1867, when they retired and opened an office as mining stock brokers.

[5]The Hale & Norcross election was the first of many conflicts between the new firm and Sharon over mines, railroad rates, lumber prices, banking, etc.

election did he learn that the stock had to be transferred in order to vote it. Flood's friends often twitted him about this in later years.[6]

At the annual election held March 10, 1869, Mackay and his friends were elected trustees, Flood was made President, and Fair superintendent. Before the election another assessment of $80,000 had been levied and in part collected, but the new trustees, believing they could make the mine pay, rescinded the assessment and returned the portion already paid.[7]

All of the members of the new firm had come to California in the early days and were men of limited education. Mackay, Fair, and O'Brien were born in Ireland, Flood, of Irish descent, was born in Brooklyn, New York. Strange to say, all were members of the Masonic Order.[8] Fair was a shrewd, closefisted businessman, a driving, indefatigable, and capable mine superintendent, and a man of considerable financial ability—but there all praise ends. Socially he was as genial as the sun, his mouth dripped honey, and his tongue was smoother than oil, but his heart was a stone and duplicity second nature. On the Comstock he was known as "Slippery Jimmy." Wealth made him avaricious and brought out his real nature. He took pride in being cleverer and trickier than other men, as appears in the documents in the Bancroft Library leading up to the flattering biography in Bancroft's "Chronicles of the Builders,"[9] that cost him $15,000. He "besmirched" all of his old partners after their unhappy relations terminated and died divorced by his good Catholic wife, complaining, "Here I am alone, bereft of friends." The "Chronicle" commented in its obituary notice: "It is said that James G. Fair did not have an intimate friend in the world."[10] The "Examiner" published devastating accounts of his life and character on December 30 and 31, 1894, the last dealing with his unhappy relations with his children.

Flood, the representative of "The Firm" in San Francisco, was a quiet, diligent, honorable man of rare business ability, who proved himself a valuable and faithful associate, although his

---

[6]James E. Walsh.

[7]Hale & Norcross annual report, March 10, 1870.

The only other company that rescinded an assessment was the Kentuck (in which Mackay was heavily interested), after the mine found its first ore.

[8]Records of the Secretary of the Grand Lodge, San Francisco.

[9]Vol. 4, pp. 209–236.

[10]*San Francisco Chronicle*, December 30, 1894.

talent for stock market operations[11] led the firm later into speculations which brought criticisms and anxieties but no profit in
the end. He was consulted on all important matters. Among
other responsibilities he played the part of peacemaker between
Mackay and Fair until his patience too was exhausted. Mackay
and Flood deeply respected each other.

O'Brien, a genial man of no particular ability, had the good
fortune to be carried into affluence by Flood, his loyal old-time
partner in the saloon business, who would "have nothing that
Billy does not share."[12]

Mackay's business and administrative abilities were now fully
developed, and in all that pertained to mining and milling on the
Comstock he had no superior. To him fell the general conduct of
affairs and supervision of the mills. In addition he acted as joint
manager of mines with Fair, although without either title or
salary. In the hands of three such attentive and businesslike
managers the affairs of the firm and the mines they came to control were conducted with the exactness of a modern manufactur-
ing enterprise.[13]

The Hale & Norcross straightway began to prosper. Every
economy consistent with good management was introduced.

The new firm soon justified itself by making the mine pay
dividends on ore that milled only $25 a ton. During the first
year (1869) dividends to the amount of $192,000 were paid. In
1870 Fair drove a crosscut 63 feet into what was thought to be
the east wall and found a new ore body, which brought the dividends for that year up to $536,000. After that year the ore on
the lower levels grew baser and smaller and production came
mainly from ore on the old upper levels in which there was little
if any profit. The ore was almost exhausted and floods of water
in the lower levels vastly increased expenses. After a final divi-
dent of $80,000 in 1872 the mine again entered the assessment
list.

---

[11]James R. Keene, that prince of manipulators, said that Flood had more
natural talent for stock operations than any other man he had ever known. It
was one of Keene's favorite sayings that "Any man who deals in stocks for any
length of time goes broke." He came near proving that in his own case, for
only a fraction of his many great fortunes remained at the time of his death.

[12]Mackay's devotion to Jack O'Brien is another example of those pioneer
friendships.

[13]*New York Tribune*, August 27, 1875. An examination of the correspendence
and the books of the Con. Virginia confirms this. Their purchases of machinery,
materials, and supplies amounted to many millions, and while they dealt fairly
they insisted upon a dollar's worth for every dollar spent.

Flushed with success the new firm then looked for another mine. The recent discovery of the Crown Point bonanza led men to believe that other great ore bodies would be found below, and Mackay and his associates decided to try to find one in the Con. Virginia, which had been a failure. That story will be told after the Crown Point bonanza.

THE VIRGINIA AND GOLD HILL WATER COMPANY

Mackay now had associates as ambitious as himself. The next move showed their forward spirit. No sooner had the Hale & Norcross declared its first dividend, in the fall of 1869, than the new firm bought Sharon's interest in the Virginia & Gold Hill Water Company, which controlled the poor and inadequate water supply of the Comstock; the remaining stock being held chiefly by their friends, W. S. Hobart and Johnny Skae.[14]   That investment was made on faith in the future of the camp, which was then in the midst of a depression. Fortunately, the Lode took on new life following the discovery of the Crown Point-Belcher bonanza at the end of 1870, and the energetic owners of the Water Company dared to attack the great problem of bringing a new supply from the tops of the Sierras.[15]   Not only did the undertaking call for expenditure of more than a million dollars, but it had been declared impracticable from an engineering standpoint, on the theory that no pipe could be made that would stand a vertical pressure of 1,720 feet.   The pipe was to be laid in the form of an inverted siphon, with the long arm running up the east side of the Sierras, the other reaching to the summit of the Virginia Range, with the depressed curve crossing the upper end of Washoe Valley.   When some one raised the old question whether a pipe would stand that enormous pressure, Flood characteristically replied, "Everything can be done nowadays; the

---

[14]They organized a California corporation of the same name, the Virginia & Gold Hill Water Company, and elected Walter S. Dean, W. S. Hobart, John Skae, John W. Mackay, James G. Fair, James C. Flood, and W. S. O'Brien as directors. Hobart, who was elected president, became very wealthy through his interests in mines, mills, water and lumber.

[15]The first pipe is described with all of the mechanical details, including diagrams, in the *Mining and Scientific Press* of December 12, 1873.

Dan DeQuille's *Big Bonanza*, pp. 233–236 (1876), describes the first two pipes; and his *History of the Comstock Mines*, pp. 64–68 (1889) tells the story of all three.

John D. Galloway, a San Francisco civil engineer who lived on the Comstock as a boy, describes this and other engineering projects in *Early Engineering Works Contributory to the Comstock*, published by the Nevada Bureau of Mines, Mackay School of Mines, University of Nevada, in 1947. (Editors, 1966.)

only question is—will it pay ?"[16] To the great credit of Hermann Schussler, who was for a lifetime chief engineer of the Spring Valley Water Works in San Francisco, the engineering difficulties were overcome, and thereafter a stream of unexcelled water sweetened the lives of the people. The day that the water was turned into the pipe, July 29, 1873, the people of the Comstock were in a fever of excitement and anticipation, and when the water actually flowed into Bullion Ravine "The crowd was as wild with joy as were the Israelites when Moses smote the rock. Cannons were fired, rockets sent up, and bands of music paraded the streets." Then leaks occurred in the pipe, the water was turned off, and people's hearts sank. For a few days it was feared that the pipe would not hold. Then, on August 8, the flow came on again uninterrupted.[17]

---

[16]Lord's *Comstock Mining and Miners*, p. 323 (1883).

A year earlier, in 1870, an iron pipe 30 inches in diameter and 14,000 feet long was manufactured and laid by the Risden Iron Works, under the direction of Hermann Schussler, for the Spring Valley hydraulic mine at Cherokee, Butte County, California. At one point it crossed a gorge where the vertical pressure was 910 feet. (*U. S. Mineral Resources* for 1872, pp. 392, 410.)

[17]Lord's *Comstock Mining and Miners*, pp. 321–325, 332, 333 ; Shinn's *Story of the Mine*, pp. 63–67 (1896).

CHAPTER XIV

1869 A Discouraging Year—The Yellow Jacket Fire—Sharon Builds the V. and T. Railroad.

The year 1869 opened with fair promise. The Comstock had become accustomed to moderate prosperity and accepted it thankfully. The boundless enthusiasm of 1863 was only a memory. Many of the lesser mines had closed down, and the leaders, with the exception of the few that had ore, were operating with reduced forces. Only the Savage, Chollar-Potosi, Yellow Jacket, Crown Point, and the Kentuck were paying dividends, and the limits of their ore bodies were known. The Sierra Nevada also was paying a trifling sum from very low grade ore. It was a memorable year for Sutro, for he was enabled after four years of disappointments to start work on his tunnel. The great hope of the year was the construction of the Virginia & Truckee Railroad from Gold Hill to the Carson River mills and to Carson City, which would reduce the cost of hauling ore and lumber and firewood by wagons and encourage the extraction of low-grade ores.

Then, on a pleasant April morning fell the greatest disaster in the history of the Lode—the Yellow Jacket fire, in which thirty-seven men were trapped underground and lost their lives.[1] Three adjoining mines (the Crown Point, the Kentuck, and the Yellow Jacket) were working on the same east ore bodies from the 600 to the 900-foot levels and their extensive stopes were a maze of large resinous pine timbers. The fire, of unknown origin, started on the 800-foot level of the Yellow Jacket and had been burning for several hours without knowledge owing to heavy doors in the drifts. As fate would have it, when the men on the morning shift were lowered down the shafts a mass of charred timbers in the stopes broke under the weight of the roof, sending a blast of deadly gas and smoke through the workings of the three mines. A few were hoisted back, many were suffocated, and others burned. The scenes about the mouths of the shafts were heartrending during the three days in which heroic efforts were made to reach the remaining men. When it became clear that all below were dead and that not even their bodies could be recovered at that time, the shafts were sealed and a pall settled over

---

[1]Another fire in the Yellow Jacket on September 20, 1873, on the 1,300-foot level, took the lives of six men.

the region.[2] The last descent into the Crown Point prior to the second sealing of the shaft was made on April 12 (the fire occurred on the 7th) by Superintendent Jones and a young man who tried to connect a pipe with the blower tube. Foul air drove them out after fifteen minutes without making the connection.[3] After the shafts were sealed large volumes of steam were forced into the workings to check the fire. Those mines, which had been among the most productive on the Lode, were practically ruined. The caved stopes smouldered for months and yielded but little good ore afterward. Instead of paying dividends all three mines began to levy assessments.

Two new discoveries were made during the latter part of the year: a body of fair ore in the Hale & Norcross, under the management of the new firm of Mackay, Fair, Flood, and O'Brien, and a promising ore body in the Yellow Jacket to the north of the fire area. Stocks continued to decline and the year ended in discouragement, with a total production of only $7,500,000—the lowest since 1862.

SHARON BUILDS THE VIRGINIA AND TRUCKEE RAILROAD

The daring and strategy of Ralston and Sharon were never better illustrated than in the building of the Virginia & Truckee Railroad from Gold Hill to the Carson River mills and to Carson City during the critical year of 1869. Those mills, driven by water power, could reduce ore at one half the cost to nearby mills driven by steam with expensive wood for fuel, but hauling to the river by wagon cost $4 a ton, and was subject to the condition of the roads in winter and early spring. A railroad hauling ore at $2 a ton would make a large volume of low-grade ores available and greatly profit the Bank Crowd's river mills. To build the road at the expense of others was the next thought. A publicity campaign was soon in full cry and the people of the region enthusiastically responded. The counties of Storey and Ormsby bonded themselves for $500,000, which was made a gift to the enterprise on the promise of largely increased taxes on the railroad; the mines, which were largely controlled by Sharon, contributed $700,000 more, part of which was to be returned in

---

[2]The details of the fire and of the attempts at rescue are set forth on pages 269–277 of Lord's *Comstock Mining and Miners* and pages 176–183 of Dan DeQuille's *Big Bonanza*.

[3]*Gold Hill News*, April 7–15, 1869.

A slight earthquake occurred on December 26, 1869. The miners underground felt it and came to the surface. No damage was done.

freight. Sharon called in I. E. James, the outstanding Comstock engineer: "James, can you run a railroad from Virginia City to the Carson River?" "Yes." "Do it at once."

Ground was broken for the road on February 18, 1869, the work was pressed with great skill and energy, and on November 12 of the same year "the crookedest railway in the United States"[4] was completed to Carson City. "In this way Sharon built the road without putting his hand into his own pocket for a cent," so Dan DeQuille wrote in 1875.[5] Dan was a reporter on the "Enterprise" when he wrote this, and Sharon was his employer. Sharon himself boasted, "I built that road without it costing me a dollar."[6] When the question of taxing the railroad arose, its financial statements exhibited an amazing increase in the cost of construction and equipment. The road managed to pay little in taxes, and the counties struggled for years with their bond issues.[7]

The road was the property of Ralston, Mills, and Sharon until Ralston's death in 1875, after which the others became the sole owners.

Although its rates were high, the road was of great benefit to the Comstock region. For years it was the most profitable railroad in the United States for its length—21 miles to Carson City, 52 miles to Reno. Connection with the Central Pacific at Reno was not made until August 24, 1872, four years after the Central Pacific had reached that point, which until then had been a ranch owned by M. C. Lake, who also had a wooden toll bridge across the Truckee River.[8] The first Central Pacific train reached Reno on June 18, 1868, and the rails joined those of the Union Pacific at Promentory, Utah, on May 13, 1869.

Reno during those years was the busiest little town in Nevada; a dozen stages and hundreds of freight wagons radiated in all directions. Then, almost overnight, the Virginia and Truckee Railroad arrived and diverted the traffic, and Reno settled down

---

[4]That peculiar adjective 'crookedest' was appropriate to the road's torturous course and to the manner in which it was financed.

[5]*Big Bonanza*, p. 522 (1876).

[6]Shuck's *History of the Bench and Bar*, p. 486 (1901).

[7]Thompson & West *History of Nevada*, pp. 280–282 (1881); Shinn's *Story of the Mine*, pp. 166–169 (1896); Lord's *Comstock Mining and Miners*, pp. 251–256 (1883).

See also, John D. Galloway's "Memorandum" on the construction of the V. & T. Railroad; one copy at the Bancroft Library, another at the Mackay School of Mines.

[8]Lake built the Lake House where the Riverside Hotel now stands.

as a small railroad town. No longer was the Geiger Grade crowded with many-muled freight wagons; no longer did fine stages drawn by handsome horses and crowded with passengers dash along C Street—to the great joy of the small boys and the delight of the multitude.

CHAPTER XV

The Gloomy Year of 1870—The Crown Point Revival in 1871—
The Boom of 1872—Sharon-Jones Contest for Senator.

The year 1870 was the darkest year in the history of the Comstock up to that time, although the production of $8,319,698 slightly exceeded that of 1869 owing to a substantial increase from the Chollar-Potosi and the Yellow Jacket, which had found new ore bodies of limited extent.[1]

Extensive development work on all of the lower levels had been discouraging. In the Gould & Curry and the Chollar-Potosi the walls of the Lode had almost come together. Besides, nearly all of the mines were struggling with increasing flows of water. Most of the upper bonanzas had played out at 500 feet. The only mines whose ore had extended to the 1000-foot level were the Savage, the Hale & Norcross, and the Yellow Jacket, and that was becoming narrower and poorer. The most hopeful view was that the mines had reached a barren zone and that other ore bodies would be found below. A vain hope it proved in after years, for the Lode continued barren to the deepest levels from the Gould & Curry to the Crown Point.

Stockholders were becoming discouraged and allowing their shares to be sold for assessments, which placed the burden upon those in control—chiefly Sharon and his associates. Stocks had fallen so low that the market value of a number of the prominent mines was far less than the cost of the machinery. This was especially true of the Crown Point and the Belcher. The former, with 12,000 shares, sold for $2 a share in November 1870. The Belcher, with 10,400 shares, was selling at $1; its shaft was full of water to the 800-foot level, and the mine was all but closed down. The total market value of the stocks of all of the leading mines on the Lode in November was less than $4,000,000. The Consolidated Virginia, after a last despairing effort on the part of the old stockholders, ceased to operate and the stock fell to $1 a share.

---

[1] The forty mines operating on the Comstock lode produced 2,000,000 tons of ore during the years 1866–1870, inclusive, yielding $60,000,000, or $30 a ton. The dividends amounted to $12,184,920, and the assessments totaled $7,253,634, leaving a profit of $5,031,866. The only highly profitable period was during the '70s when the Crown Point and Con. Virginia bonanzas paid $100,000,000 in dividends. But even then the heavy assessments levied by the other mines reduced the net profit of the mines on the main Lode to $55,000,000.

A number of the leading mining men and many miners left the Comstock in 1870 for the great silver camp of Pioche and the still more productive silver-lead camp of Eureka, both of which were just coming into prominence.

Ralston and Sharon were almost panic-stricken. They had staked everything on the future of the Lode—not only their newly made fortunes, but the solvency of the Bank. Lord tells of the perilous condition at that time:

> Mr. Sharon has said that the amount invested by the Bank of California at one time in the mines, mills, and towns directly dependent upon the continued productiveness of the Comstock Lode was $3,000,000. The whole capital of the bank in 1870 was $5,000,000, and, though the great moneyed institution of the Pacific Coast, the loss of this investment, or even a popular dread of such a calamity, would have endangered its very existence, and certainly have crippled it for a time. Only the few directly acquainted with the condition of the bank will ever realize the anxieties which beset its management at the close of the year 1870.[2]

Sharon's biography states that "Ralston revealed to Sharon the peril which threatened him and the bank" unless the Comstock produced another bonanza. It was a narrow escape for the Crown Point was about to be closed down when the discovery was made.[3] If that bonanza had not been found it is probable that pumping would have ceased along the Lode and mining thereafter confined to the extraction of the remaining ore on the upper levels. The Sutro Tunnel, which was started in a small way in October 1869, would have been abandoned. Only 1,750 feet had been dug at the end of 1870, and that of small size and cheap construction. It had 18,739 feet further to go before reaching the Lode.

Perhaps years would have elapsed before renewed efforts would have been made to find ore below the 1000-foot levels, and then with doubtful success. The only two large ore bodies discovered later in the entire length of the Lode were found in unlikely places. By a freak of fortune three of the most discredited mines on the Comstock—the Belcher, the Crown Point, and the Con. Virginia—were to be its saviors. While the stock market and

[2] *Comstock Mining and Miners*, p. 279 (1883).

[3] "A year ago the Crown Point Company had concluded to shut down their mine in January, as it could only be kept working by levying assessments and the prospects ahead were decidedly gloomy." *Gold Hill News*, September 16, 1871.

the assessment system have much to answer for, it should be remembered that neither the Crown Point nor the Con. Virginia bonanza would have been discovered without their aid. While Ralston and Sharon milked the Comstock they sustained it during the lean years of 1865, 1869 and 1870. But for that support the history of the camp might be a different story.[4]

Another welcome discovery toward the end of the year 1870 was made by the Chollar-Potosi, which had the largest body of quartz on the Comstock. It had been mining a large volume of $25 ore, mill returns, during the four preceding years and paying small dividends. Then, quite unexpectedly, the rich Belvidere ore shoot was encountered near the surface, toward the south end of the mine, which yielded 70,000 tons of $50 ore and returned $1,946,637 in dividends, which exceeded all of the dividends theretofore paid. That ore body marked the end of profitable operations and the mine went on the assessment list, where it remained.[5] Isaac L. Requa,[6] the very able superintendent, also managed Sharon's mills for twenty years and superintended the great Combination third-line shaft from its inception in 1875 until pumping ceased in 1886.

THE CROWN POINT REVIVAL IN 1871

When the black year of 1870 had all but expired, J. P. Jones, superintendent of the Crown Point, reported a discovery on the 1100-foot level. It was neither large nor rich but gave promise of greater things. Raymond, in his 1873 report, says the management "stumbled upon it" south and east of the point where

---

[4]Lord suggested that "if the managers had the lion's share of the profits they had also the lion's share of the risk and labor." *Comstock Mining and Miners*, p. 331 (1883). See also, Thompson & West *History of Nevada*, p. 594 (1881).

[5]The surface of the Chollar-Potosi is now gashed by a great open cavern caused by the extraction of about 2,000,000 tons of ore. This was largely the work of a long period of years during which the margins of the old ore bodies were mined and milled—at little or no profit.

[6]Requa, always a forward-looking man, was the inventor of the self-dumping skip, which is used in mines throughout the world. He also installed the first hydraulic pump on the Comstock, to supplement the work of the Cornish pump at the Combination Shaft. He was one of the most distinguished citizens of the Comstock for twenty-five years. His imposing figure—6 feet, 2 inches in height and 240 pounds in weight—was matched by deliberate manners and speech which never failed to carry conviction. When the Combination Shaft was closed down he retired to Oakland with a fortune and became a successful banker. His son, Mark, was a successful mining engineer. He and Fred W. Bradley started the Nevada Copper mine at Ely on its great career in 1904.

the earlier east ore bodies had died out on the 860-foot level.[7] Sharon and his associates, including Hayward, had been in control of both Crown Point and Belcher for years, although without holding a majority of the stock in either mine after assessments began to be levied. Crown Point stock rose quietly from $3 a share on November 19 to $16 on December 10. The market price of Belcher, meantime, had increased to $7.50 a share. Development work proceeded slowly for a time and the reports were not very encouraging, due no doubt to the plans of Hayward and Jones who were aiming to secure control.

"Jones' sick baby" was given as one of the excuses, which later became a byword on "The Street" in San Francisco. Jones, who was without means, induced some San Francisco speculators to buy stock for him "upon his agreement to bear all losses in consideration of one half of the possible profits." Later, he advised these men to sell, "although assuring them of his firm belief in the mine. * * * They regarded his story as a lame pretense," and began to sell short, which later cost them dear.[8]

The ore had been found only 200 feet north of the Belcher line, and, as it lay on the footwall of the Lode, there was every probability that it would extend into the Belcher. That stock also began to rise. Crown Point stock advanced slowly, with wide variations, during the first half of 1871, evidently manipulated up and down, reaching $300 in June. Belcher went through the same performance, and reached $240 by the end of that month.

Hayward, representing himself and Jones in San Francisco, began to buy Crown Point stock from the time of the discovery. Lord says: "His purchases were made so rapidly and shrewdly that he obtained 5,000 shares, nearly half of the entire stock in the company, at prices averaging less than $5 per share." That statement is questionable. It is highly improbable that he acquired that amount of stock at such low prices. The brokers on the San Francisco Exchange were quick to note a demand and boost the price. The stock rose beyond $5 immediately after the discovery and was selling at $18 within three weeks. A few months later Hayward bought 1,000 shares from Charles B. Low at prices ranging from $90 to $180, which assured control of the mine.

---

[7] It was said that credit for the discovery was due W. H. "Hank" Smith, the able and popular foreman of the Crown Point. Smith followed Sharon, became superintendent of the Belcher and other Gold Hill mines, and had a distinguished career on the Comstock and later in Utah.

[8] Lord's *Comstock Mining and Miners*, pp. 282, 283 (1883) ; Marye's *From '49 to '83*, pp. 125, 126 (1923).

Sharon appears to have been slow to learn what was going on. It is probable that he could not conceive that his associate, Hayward, and his superintendent, Jones, would play him false. Instances of that kind were rare. While warring groups of speculators fought each other with all of the means at their command—deceit being the chief weapon—they played fair among themselves. It is quite probable that both Hayward and Jones had become rebellious over little Sharon's autocratic manners and methods.

Meanwhile Sharon had been manipulating Belcher and gathering in all available stock at comparatively small prices. His biography[9] states that he and Ralston secured almost all of the stock in the Belcher at $1 a share. They must have bought it from the treasury, to which it had returned for nonpayment of assessments, a favorite trick of those in control when a boom started.

Sharon was bitter when he learned that Jones and Hayward had the control of Crown Point, but made the best of the situation: "He proposed to sell to Hayward and his friends all the shares of himself and friends in the Crown Point mine, at the market price, on condition that Hayward and his friends would sell to them all of their interests in the Belcher."[10] The offer was accepted and the transaction closed. Lord states on the authority of Sharon that Sharon sold his 4,100 shares of Crown Point to Hayward on June 7, 1871, for $1,400,000, which would be at the rate of $341 a share. This was by far the largest single transaction in Comstock stocks up to that time.

The boundary line between the two mines passed downward through the middle of the bonanza, dividing it into two nearly equal parts. The Belcher's portion proved richer and more productive so that the trade eventually favored Sharon and his associates Ralston and Mills. In addition to the lion's share of the dividends, each group made private milling profits running into the millions.[11]

It has been estimated that the Comstock profits of Ralston, Sharon, and Mills amounted in all to about $20,000,000, although

---

[9] *Chronicles of the Builders*, Vol. 4, p. 41.

[10] King's *History of the Stock Exchange*, p. 47. See also, Lord's *Comstock Mining and Miners*, pp. 282–284; Marye's *From '49 to '83*, p. 125 (1923); Shinn's *Story of the Mine*, pp. 170, 171 (1896).

[11] A considerable number of Crown Point stockholders organized in May 1877, and protested vigorously against the practice of the management in milling ores in their private mills instead of the company's mill. *Mining and Scientific Press*, May 19, 1877.

Ralston's speculations and widespread business enterprises broke him at the very summit of success. Sharon's biography says that "before the year 1875, the Union Mill and Mining Company had netted Mills over $2,000,000, and Ralston and Sharon over $4,000,000 each." Their other dividends and profits could not have been less than $10,000,000. Sharon said to George T. Marye, Sr., about 1874, that he, Sharon was the second richest man in California; his associate D. O. Mills being the first.[12] When Sarah Althea Hill sued Sharon for divorce, based on a marriage contract, in January 1884, after he had brought suit to have the alleged contract invalidated, she stated his fortune to be $15,000,000. His sworn answer denied that he was worth to exceed $5,000,000, which may be accepted as a conventional denial.[13]

The Comstock knew only Sharon and hated him for his ruthless methods and dictatorial manner; yet he was cock-of-the-walk only on the Comstock. During all of the early years he was subordinate to Ralston and never made an important move without consulting him. After the Crown Point-Belcher bonanza had made them all rich beyond their dreams, Sharon became more independent and, when the Bank of California suspended on August 26, 1875, followed immediately by Ralston's tragic death, he took the lead, and, with the aid of sixty-three loyal citizens, restored the Bank's capital and reopened its doors within six weeks.

THE BOOM OF 1872

It is a feature of stock market operations that a leader carries the market up or down. That was the invariable rule with Comstocks. Speculators, rich and poor, were always eager to buy on a rising market. If their means were small they bought the cheaper stocks, which were certain to advance. A discovery in one mine raised the prices of all of the stocks on the board, often higher proportionately than that of the mine in which ore had been found. This was largely due to manipulation. When stocks were low and mines perhaps closed down, most of the stock reverted to the treasury for nonpayment of assessments. Then, on the first indication of a boom, those in control would purchase

---

[12]Marye's *From '49 to '83 in California and Nevada*, p. 126 (1923).
[13]The *San Francisco Chronicle* of January 4, 1875, boasting of the city's rich men, credits the following with fortunes of $5,000,000: "Lick, Latham, Sharon, Hayward, Reese, Mills, Baldwin, Lux, Miller, Jones, Ralston and Stanford." The wealth of some of those men is understated.

the treasury stock at the market price and proceed to boost the stock by wash sales and otherwise.

The so-called "Boom of 1872," when 150 stocks on the board made such remarkable advances from January to May 1872, was a man-made affair, manipulated by Alvinza Hayward, who deliberately and openly boosted Savage stock from $62 to $725 a share on the pretense that a rich discovery had been made in the mine. The market followed as a matter of course.

The stocks of the Crown Point and the Belcher had hung around $300 a share during the latter half of 1871, and were worth it. All of the others on the exchange had advanced with them, but there was no boom until after the first of January 1872. The developments in Crown Point and Belcher by that time were so favorable that all of the other stocks began to increase rapidly in price.

It was then that Hayward launched his spectacular boom in Savage, partly for the purpose of furthering the senatorial aspirations of his partner Jones. They had taken control of the Savage from Sharon at the annual meeting in July 1871, in order to get its ore for their mills and to display their newly acquired power.

Toward the end of January 1872, Hayward gave an unlimited order to buy Savage stock, which was then $62 a share. Next the miners were confined, and the public denied entrance to the mine. No one was permitted to see or tell of a rich strike— which had not been made.

The whole market went up with a rush. Savage rose from $62 on February 1 to $310 on the 8th, fell to $230 the middle of March, rose to $460 on April 17, and to $725 on April 25. At that price the mine was selling on a basis of $12,400,000, which was almost as much as Crown Point and Belcher were worth.

Crown Point rose to $800 on February 1, then hung around $770 until the middle of March, rose to $1,250 by April 17, to $1,700 on the 25th, and to $1,825 on May 5. Belcher followed along with Crown Point, and reached the top, $1,525, on April 25. The San Francisco Bulletin of May 7, 1872, reported:

> The excitement in mining stocks and mining claims during the past few months had been without a precedent in the history of our mines. Mining incorporations have been multiplied like the leaves of autumn. The capital of existing incorporations has been increased in the most lavish manner. Prices have gone up like a rocket, and in some cases have reached altitudes never

dreamed of even by the most enthusiastic. Yet it is noteworthy that out of 150 claims offered to the public through the stock boards, only four are paying dividends. These are the Belcher, Crown Point, and two companies at Pioche, Nevada.

The stocks in the 150 mines listed on the exchange (which included many scattered all over the West), had increased in value from $17,000,000 in January to $81,000,000 on the 5th of May. A crash was inevitable and was hastened by the moves of Sharon who was a rival candidate against Jones for a seat in the United States Senate. On May 8, Sharon let it be known that he had information that J. P. Jones had been instrumental in setting the Yellow Jacket fire in April 1869, in order to break the market at that time. The charge was baseless and almost absurd, although Sharon had the affidavits of several irresponsible men. The market had reached tottering heights and the sensation brought it down. Stocks dropped 30 to 40 percent. The panic that followed shook San Francisco like an earthquake; speculators saw their fortunes crumble. Crown Point fell from $1,825 on May 5 to $1,659 on the 8th, and to $1,000 a few days later. The Jones crowd raided Belcher, which fell from $1,400 to $750, but recovered shortly to $1,000.

One of Lord's best stories is spoiled by a little fact. He says that because of Sharon's charge against Jones, Crown Point shares fell to one eighteenth of their former price and hurried on a general fall in mining stocks, overlooking the fact that on May 15 the capital stock of Crown Point was increased from 12,000 to 100,000 shares (8⅓ for 1), which left the price $100 a share. A few weeks later it rose to $135. Belcher was increased from 10,400 to 104,000 shares on August 1, 1872, and fell to $108 a share, which led innocent writers to comment on the extraordinary decline in prices.

The San Francisco Chronicle on May 19 says the financial wreck of the city is complete, and that "to Savage more than any other mine the wild furors can be traced, which, like a whirlpool, drew almost everybody in the vortex of speculation."

About the first of September 1872, a drift in the Con. Virginia crossed a fissure containing low-grade ore, giving strength to the market, which continued to advance for over two years as developments in the Con. Virginia became more and more favorable and the Crown Point and the Belcher were declaring millions in dividends.

SHARON-JONES CONTEST FOR SENATOR

Sharon suffered his first defeat when Mackay and his associates took control of the Hale & Norcross in 1869. His supremacy was first challenged in 1871, when Hayward and Jones covertly secured control of the Crown Point soon after the bonanza was discovered. Five years later, the "Bonanza Firm" or the "Bonanza Crowd," as it was sometimes called, completed his dethronement. The "Boom of 1872" made Sharon a millionaire several times over, but he was not content. He was a vain little man, not seeking popularity but fond of show and coveting power. A seat in the United States Senate would gratify both. The only man in his way was John P. Jones, who less than two years earlier had been one of his compliant superintendents.

Jones, a genial man and a born politician, had a meteoric rise in Nevada. He was a large full-bodied man with a long chin beard and a benevolent countenance. A Welshman, born on the English border, he arrived in California in 1850; mined in several camps until 1852 when he settled at Weaverville, Trinity County. There he followed public life, serving successively as Justice of the Peace, Deputy Sheriff, Sheriff, and State Senator.[14] In the fall of 1867 he was a candidate for the office of Lieutenant Governor of California. That year the Democratic ticket prevailed, and Jones was left high and dry—defeated, broke, and discouraged, and about to depart for the East, as he himself said. At this juncture, in November 1867, at the behest of his friend Alvinza Hayward, he was sent to the Comstock to become superintendent of the Kentuck, which Sharon, Hayward, and associates had just acquired. Jones was evidently appointed for his diplomatic qualities, for he had never mined in Nevada and had followed political life in the main. It was not uncommon for such appointments to be made. Capable mine foremen attended to the details. When the editor of the Gold Hill News heard of the appointment, he was indignant: "There is no better man in the Comstock than John D. Winters, the present superintendent."[15] In 1868 Jones was made superintendent of the Crown Point, also controlled by Sharon and associates. The two succeeding years brought little comfort to Jones. The ore continued to fail and the mine was about to close down when the discovery was made. Then, almost overnight, he became a millionaire.

---

[14]*Gold Hill News*, August 29, 1873—facts evidently supplied by Jones.
[15]*Gold Hill News*, November 30, 1867.

Sharon was now to swallow another bitter pill. Jones, a seasoned politician, had himself called "The Commoner," spent his newly won wealth regally, and was triumphantly elected in January 1873[16]—an office which he filled with credit to himself and his State for thirty years. Senator Jones was chiefly distinguished for his eloquent advocacy of silver and as the best story teller and poker player in the Senate.

Ex-Governor James W. Nye, the incumbent Senator, who had served Nevada long and well, had no chance in that race.[17]

Harry M. Gorham, nephew to Jones, says in his interesting little volume "My Memories of the Comstock" (1939): "When Jones went to the Senate he took off a balance sheet, and he was worth $8,000,000."

The aftermath of the Hayward-Jones friendship, according to the Virginia Evening Chronicle of December 12, 1874, was a quarrel over the Crown Point, in which Jones prevailed. "Now Jones and Hayward are at swords' points. They are even more bitter in their hatred of each other than Sharon and Jones ever were."

Newly created Comstock millionaires, with the notable exception of Mackay, almost invariably aspired to a seat in the U. S. Senate, and the campaigns of Jones in 1872, Sharon in 1874, and that of James G. Fair in 1880, were said to be characterized by "a saturnalia of corruption."[18]

---

[16]Thompson & West *History of Nevada*, pp. 91, 92 (1881).

[17]Nye's pleasant wit and readiness at repartee were of no avail when he was made the butt of one of the famous witticisms of early days. As Governor he had secured an appropriation of $75,000 to build a dam and a sawmill to cut lumber for the Piutes on the reservation. The money was all used up on the dam whereupon the irrepressible W. J. Forbes wrote that "Governor Nye has a dam by a mill site but no mill by a dam site."

[18]Davis' *History of Nevada*, pp. 421–423 (1913). "The 'Battle of the Money Bags' for Senatorial Honors," it is termed in Thompson & West *History of Nevada*, p. 92 (1881). Nevada was often characterized as "The Rotten Borough."

Sutro was also a candidate, as he was in every senatorial election thereafter up to and including 1880; and each time he failed to receive a single vote in the Legislature.

SECTION THROUGH BELCHER MINE

(See bottom opposite page)

CHAPTER XVI

The Crown Point-Belcher Bonanza—The Gold to Silver Ratio— The Silver Question.

The Crown-Point Belcher bonanza was an ideal ore body. It was fairly uniform in value, easily mined, remarkably free from base metals, and, unlike the other major ore bodies, it lay upon the footwall of the Lode, which at that point had a dip of about forty degrees. The ore extended from the 900- to the 1500-foot levels in both mines, and was widest and richest on the 1300 level where it had a length of 775 feet, and, in the Belcher, a width of 120 feet measured on the horizontal. At the 1300-foot level the ore split horizontally into two bodies, like a fish with two flat tails, one branch continuing down on the footwall, the other descending at a slighter dip.[1] In the Crown Point a narrow parallel ore body was found lying about 40 feet east of the main body. This extended from the 1200- to the 1400-foot levels and produced considerable ore.

Superintendent J. P. Jones, in his report for the year ending May 1, 1873 (the first since 1870),[2] told of the ore on the different levels of the Crown Point (exclusive of an equal portion in the Belcher), and continued: "It thus appears that the ore body has steadily increased in length, width, and richness as we have descended upon it, and there is every indication of its continuing to do so. * * * It is fair to presume that we have passed below the range of surface disturbance, and that the vein will penetrate the earth in its present shape to an indefinite depth." But his hopes were soon to be shattered, for the ore body

---

[1] A cross-section of the ore body is shown on plate 7 of the Atlas which accompanies Becker's *Geology of the Comstock Lode* (1882).

[2] He prefaces his report for 1873 with the statement: "The last general report submitted by the superintendent was dated May 1, 1870. At that time the Crown Point mine was yielding nothing." He does not make any explanation for the failure to report to his stockholders in 1871 and 1872. The report of 1873 covers all three years. (It is printed in *U. S. Mineral Resources* for 1873, pp. 177–186.)

CROSS SECTION THROUGH CROWN POINT-BELCHER BONANZA—The footwall is composed of metamorphic slates, unlike the Virginia City section where it is diorite. The hanging wall is made up of three separate flows—the upper of augite andesite, the second of hornblende andesite, and the lower of diabase. The ore bodies, upper and lower, are in black. The small ore body near the surface was worked out by the Belcher in 1863, 1864, 1865—Scale approximately 880 feet to 1 inch. (From plate 7 accompanying Becker's report.)

contracted somewhat on the 1400-foot level, where the values were lower, and practically terminated on the 1500-foot level.

That wise mining engineer, Rossiter W. Raymond, in his U. S. Mines and Mining Report for 1873, written before the ore body was fully developed, warns the Comstock operators not to expect the ore to continue in depth:

> I do not doubt that the present year, while it cannot exhaust the great ore body from which the Crown Point and Belcher have obtained so much profit, and the proprietors of other mines so much hope, will nevertheless reveal more clearly than they are now known the limits of that body or of its richest mass. Whoever believes that these mines have now at last entered upon a solid and continuous body, extending indefinitely in depth, and precluding for the future the necessity of explorations, will find himself mistaken.[3]

The stopes were so large on the 1300-foot level that square-set timbering alone would not hold up the ground without reinforcement, and both mines were required to use millions of feet of heavy timbers to fill in their square sets and build bulkheads as the ore was removed, just as the Con. Virginia and the California were compelled to do a few years later and on a larger scale.

From 1870 to 1878, when production practically ceased, the Belcher produced 684,000 tons, yielding $32,118,000, or $47 a ton, and paid $14,876,000 in dividends. Its greatest year was in 1873, when 154,664 tons yielded $10,525,000, or $69 a ton. The dividends that year totaled $6,760,000.

During the same period, 1870–1878, the Crown Point produced 725,000 tons, yielding $25,877,000, or $35.70 a ton, and paid $10,740,000 in dividends. The mine was at its best during the year ending May 1, 1874, when 145,129 tons yielded $7,307,258, or $51.11 a ton, from which $5,300,000 was paid in dividends.

That great ore body lasted only four years. By 1875 Belcher ore had fallen to $25 a ton; Crown Point to $18.44. There was

---

[3]A year earlier Raymond wrote: "Both companies are digging pell-mell to see which can produce the most in the shortest time. Belcher now produces nearly 500 tons of ore daily, and is making preparations to produce between 500 and 600 tons. Furthermore, workings of this style on the Comstock have taught us what result to expect." (*U. S. Mines and Mining* for 1872, p. 118). Raymond was critical of the Comstock practice of "gutting the mines."

no profit to the stockholders from such ore, as mining costs averaged $9.50 a ton and milling $11, in addition to a heavy burden of general expense, including that of deep mining.[4]

The Crown Point paid its last dividend in 1875 and the Belcher in 1876, and both began to levy assessments in excess of $400,000 a year in order to sink their shafts as rapidly as possible in the expectation of finding other ore bodies at greater depths. Again and again floods of hot water all but overcame them, and the difficulty of ventilating the steaming workings was almost as great. Their huge pine pump rods broke repeatedly.

No payable ore was ever found in either mine below the 1600-foot level, although both were developed to the depth of 4,200 feet on the dip of the vein by incline shafts, which was equivalent to 3,000 feet vertically.

All of the Gold Hill mines ceased pumping in March 1882, after they were flooded by a rush of hot water from the 2800-foot level of the Exchequer mine that drowned the New Yellow Jacket pumps. The water in Gold Hill then rose gradually to the Sutro Tunnel level, where it has since remained. The Crown Point, Belcher, and the Yellow Jacket then began to mill low-grade ores for the benefit of their mills.

The plat on the following page shows the extent of the stopes of the Crown Point-Belcher bonanza on the different levels from the 1000 to the 1600. The dividing line between the two mines extends down through the middle of the ore body.

The stopes indicated in the upper right-hand corner near the top represent the west-dipping vein worked out by the Crown Point, the Kentuck, and the Yellow Jacket during 1865 and 1866. That vein terminated at the depth of 400 feet. The stopes below them were on the ore bodies lying on the east side of the Lode, between the 600- and the 1000-foot levels, which were worked by the three mines from 1866 to 1869.

No ore was found at greater depth in any of those mines, although the Belcher and Crown Point incline shafts reached the vertical depth of 3,000 feet, and the Yellow Jacket vertical shaft was sunk to the depth of 3,080 feet.

(From Atlas accompanying Becker's Geology of the Comstock Lode, plate 6. Each square represents 100 feet.)

---

[4] The progress in the development of the Crown Point-Belcher bonanza is described in Raymond's *U. S. Mines and Mining* for the years 1871 to 1876 inclusive.

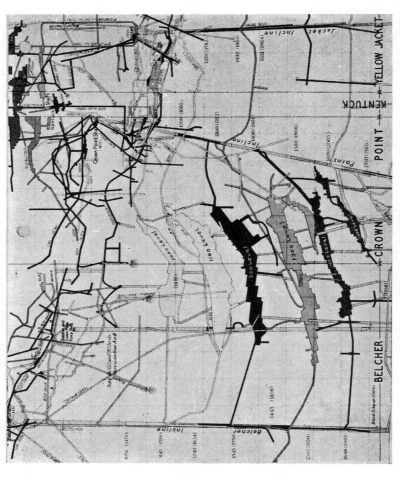

PLAN OF STOPES OF BELCHER, CROWN POINT, AND YELLOW JACKET

(Description, p. 139)

THE GOLD EQUALS OR EXCEEDS THE SILVER AT DEPTH

There was a surprising increase in the proportion of gold in the Crown Point and Con. Virginia bonanzas. In the Ophir, Gould & Curry, Savage, and Hale & Norcross bonanzas the value of the silver was almost double that of the gold, while in the two great bonanzas found below the 1000-foot level the production of gold equaled or exceeded that of the silver.

The yield of the Crown Point-Belcher bonanza from 1871 to 1878 was 54 percent gold and 46 percent silver. The production of the Crown Point's portion of the ore body was 45 percent gold, while in the Belcher, where the ore was richer, the yield was 60 percent gold. The production of the two mines that shared it from 1871 to 1878 was:

	Gold	*Silver*
Belcher	$19,142,165	$12,704,611
Crown Point	11,652,288	14,224,520
	$30,794,453	$26,929,131

When the fills and margins of the old bonanza stopes of the Belcher and the Crown Point were mined in the early '80s the production was about one half gold.

The value of the gold in the Con. Virginia bonanza, taken as a whole, was equal to that of the silver. Again the gold values were irregular, for the California's portion yielded almost 54 percent gold, while Con. Virginia ore returned only 48 percent. Such an increase in gold values in silver mines in very unusual. The gold and silver production of the two mines from 1873 to 1880 was as follows:

	Gold	*Silver*
Con. Virginia	$29,168,226	$31,959,256
California	23,395,270	20,646,106
	$52,563,496	$52,605,362

When the old stopes of the Con. Virginia and California were reopened and mined for low-grade ore, from 1884 to 1897, the yield was $9,243,787 in gold and $7,456,031 in silver. The continued decrease in the market price of silver accentuated the gold production.

Lord tabulates the output of the Lode in part, and finds the relative proportions of the silver and gold produced by all of the mines up to 1880 to be 57 to 43, estimating silver after 1872 at $1.2929 an ounce.[5] At the market or coin value after silver was demonetized on February 12, 1873, the proportions would be 55 percent silver and 45 percent gold. Nearly all of the ore mined on the Comstock after 1880 came from the fills and the margins of the old ore bodies and was low in grade. The silver content was high (with the exception of the ores from the old stopes of the Con. Virginia and Crown Point bonanzas), but that was offset by a constantly declining price, whereby the gold values equaled the silver. The total yield of the Comstock from 1859 to 1938 may be placed in value at 55 percent silver and 45 percent gold.

Comstock mines reported their silver at $1.2929 an ounce throughout nearly all of their history. When it sold at a premium, prior to 1873, the annual reports credited the excess as "premium on silver." Afterward the "discount on silver" was charged as an expense of operation. The discount amounted to approximately $20,000,000 in all, and has misled the chroniclers who accepted the production as stated in the annual reports. In this book silver is reckoned at its market value in all statements of the production of mines after its demonetization in 1873.

THE SILVER QUESTION

The great flow of gold from California in 1849 and 1850 alarmed the bankers of Europe. Holland and Belgium, in 1850, began to sell their gold and stock up with silver. Other nations followed, especially after the great gold discoveries in Australia, beginning in 1851, and silver rose in price throughout the world. The countries of Europe, with the exception of Great Britain and France, were practically on a silver basis until 1871 when

---

[5]*Comstock Mining and Miners*, p. 418 (1883) ; Becker prints the same tables on pp. 7–9.

The Crown Point-Belcher bonanzas, with its great yield and potential future, doubtless influenced the demonetization of silver. The Con. Virginia bonanza was not discovered until March 1, 1873, two weeks after the Act was passed.

A party of European capitalists, mostly French, arrived yesterday from Lake Tahoe. Today they spent several hours exploring the Belcher mine. (*Gold Hill News*, July 7, 1873). That great body of ore, of unknown possibilities, may have influenced France to suspend free coinage in 1874.

A special train from California arrived at Gold Hill on July 4, 1873, bearing W. C. Ralston, William Sharon and family, Michael Reese, Mr. Gensi (special agent of the Rothschilds), Mr. Newlands, Daniel C. Gilman (President of the University of California), Stephen Franklin (Secretary of the Bank of California), ex-Mayor Thomas H. Selby, Mr. Seward (U. S. Consul at China). The party visited the Crown Point and Belcher mines and departed for Lake Tahoe. (*Gold Hill News*, July 5, 1873.)

Germany adopted the gold standard after receiving a large amount of gold from France in payment of the war indemnity. Japan and the United States demonetized silver in 1873, and Denmark, Sweden, Norway, France, and Holland soon followed. All of those countries threw quantities of silver upon the market, with a resultant decline in price.

Silver sold at a premium from 1859 until demonetization in 1873, when it would no longer be coined free at $1.29 an ounce, $\frac{1}{16}$ the value of gold. The market price was $1.36 an ounce in 1859, from which it declined slowly to $1.32 early in 1873, although the coinage value was $1.2929 an ounce. During 1873, when silver was demonetized, the average market price continued to decline until it reached $1 in 1886. After that the decline was rapid.

The long and bitter struggle of the silverites to undo "The Crime of '73" and restore the white metal to its old-time parity of 16 to 1 with gold, was begun in 1876 and became a national issue with the defeat of Bryan for president on that platform twenty years later. Meantime, the Democrats, aided by western Senators, succeeded in passing the Bland-Allison bill in 1878, providing for the purchase in the open market and coinage of not less than 2,000,000 nor more than 4,000,00 ounces of silver per month. This was repealed by the Sherman Act of 1890, which authorized the purchase, but not the coinage, of 4,500,000 ounces of silver per month. In 1893, after the panic had set in, a Republican Congress, aided by gold Democrats, passed a bill repealing the Sherman Act, which was signed by President Cleveland, thereby leaving silver without Government support.

THE BONANZA FIRM IN 1875

JOHN W. MACKAY
JAMES C. FLOOD

JAMES G. FAIR
Wm. S. O'BRIEN

CHAPTER XVII

The Consolidated Virginia Bonanza: Early History—The Firm Takes Control—The Discovery of Low-Grade Ore—The Top of the Bonanza Encountered—Capital Stock Increased As Bonanza Develops—Dan DeQuille's First Report on the Bonanza—The California Organized and Mills Built.

EARLY HISTORY

Mackay never took credit for leading his associates from one enterprise to another, in fact, he never sought credit for anything, but from the time he got a foothold on the Comstock every step was forward until death intervened forty years later. Fair always claimed the glory for the successes of the Firm; the failures he laid to his associates. In after years when Mackay no longer spoke to him and Flood was cold, Fair used to brag: "Those lads would be in overalls but for me." It is probable that he did recommend the Hale & Norcross venture, which started the Firm on its career.

The Hale & Norcross, after three years of success, had become a problem; the ore on the lower levels was lean and base, the volume of water had largely increased, and the expense of operating was almost double. Our ambitious Firm had made nearly $1,000,000 in profits from that first venture, and, encouraged by the discovery of the bonanza on the lower levels of the Crown Point, decided to take a gamble on finding an ore body at greater depth in that discredited section of the Lode lying between the Ophir and the Gould & Curry bonanzas, which had been the graveyard of many hopes. Here was a chance to make a new mine—just such a venture as Mackay welcomed, now that he could afford it. The Kentuck and the Hale & Norcross had made him almost a millionaire.[1]

In the early '60s this was the most highly prized of any undeveloped portion of the Comstock. The Lode there was 1,000 feet wide at the surface and filled with one of the largest bodies of quartz in the region. The fact that it was almost barren of silver and gold at the surface was no sign that rich deposits would not be found below. Other ore bodies had occurred in similar masses

---

[1]James E. Walsh. Walsh was employed in the San Francisco office of the Firm from boyhood, and managed the Flood Estate until his death a few years ago.

of nearly barren quartz. Shortly after the Ophir bonanza was developed, five of the six small claims covering that section were acquired by Californians and conveyed to corporations, whose stocks sold on the exchanges at high prices for several years.

The Central, which adjoined the Ophir and had had the southerly edge of the Ophir bonanza, had 150 feet of the Lode; the adjoining (old) California 300 feet; Central No. 2, 100 feet; Kinney (not incorporated), 50 feet, White & Murphy, 210 feet, and the Sides, 500 feet—1,310 feet in all.[2] All of those companies energetically developed their several properties for four or five years by means of shafts and tunnels. The V-shaped upper portion of the Lode in that section was literally honeycombed with mine workings to the depth of 500 feet, without yielding a ton of profitable ore except the Central, which soon worked out its small segment of the Ophir bonanza. The California found a small body of low-grade complex ore near the surface, in which there was no profit; otherwise the workings in the several mines disclosed nothing but almost barren quartz and silicified porphyry. No other portion of the Lode was so disappointing in the early '60s.

The final early effort was made in 1864, by the driving of the Latrobe Tunnel, 2,800 feet in length,[3] which encountered the Lode on its dip 700 feet below the croppings and drained all of the mines in that section. Drifts were extended for 1,000 feet north and south along the Lode, which had narrowed to 200 feet and remained almost barren. Work then ceased in those properties for nearly five years.

The general revival of mining in 1867 led the large stockholders in the Sides, the White & Murphy, and the (old) California to consolidate those companies in order to sink a new vertical second-line shaft, 1,500 feet east of the croppings, in the hope of finding ore at greater depth. Accordingly, the Consolidated Virginia Mining Company was organized June 7, 1867, with 1,160 shares of the par value of $2,000 each. The incorporators were William E. Barrow, Frank Livingston, A. P. Crittenden, Solomon Heydenfeldt, and Louis Sloss. The purpose was to include five of the little properties, owning 1,160 feet of the Lode, but the Central No. 2 and the Kinney failed to join the organization, which left the Con. Virginia with 1,010 feet.

---

[2]Lord tells of the location of these claims and of the adjustments of boundaries by the locators. *Comstock Mining and Miners*, pp. 46–48 (1883).

[3]The Latrobe Tunnel passed over the largest part of the Con. Virginia bonanza, 600 feet above the top of the ore body, of which there was no sign in the tunnel.

Difficulties of one kind and another interposed and the shaft was not started until two years later. The ambitious plan was to sink the shaft vertically to the depth of 1,500 feet, at which point it was expected to intersect the Lode on its easterly dip. If that had been done the resulting failure would have changed the subsequent history of the Comstock. The Lode, which was 1,000 feet wide at the surface, narrowed rapidly with depth and closed to a mere contact between walls at 900 feet. Nor was any ore ever found below that point in the Lode itself. No one dreamed of a great body of ore standing in a rift in the hanging wall 700 feet east of the Lode on the 1200-foot level. If the shaft had intersected the Lode at the depth of 1,500 feet and found it to consist of clay and fragments of porphyry, that section of the Lode would be believed barren, and it is more than probable that the bonanza would have remained undiscovered to this day. That great body of ore had one narrow escape after another.

Times were hard during 1869 and 1870 when the Consolidated Virginia was sinking its new shaft, and when it reached the 500-foot level the managers gave up the idea of sinking deeper and decided to drive a 900-foot crosscut west to the Lode, which was encountered 200 feet below the Latrobe Tunnel. Again long drifts were extended north and south, and again nothing of value was found. The Lode had narrowed on that level to 150 feet of practically barren quartz and porphyry. All work then ceased, and the stock, which had been increased to 11,600 shares, fell to $1 a share in July 1870. The stockholders had paid $161,349.41 in assessments for the last venture and were utterly discouraged. During the ten preceding years those six little mines had expended not less than $1,000,000 without any return.

THE FIRM TAKES CONTROL

The Consolidated Virginia in 1871 was just "a good mining gamble," as Lord put it. Mackay and his associates fully understood that if the Lode continued barren at greater depths the enterprise would end in failure. As fortune would have it, no similar mining venture in history ever brought such rich and unexpected returns. They hoped to find an ore body, and would have been content with a fair measure of success.

The stocks of all of the mines in the region rose in price after the discovery of the Crown Point bonanza. Con. Virginia, which had sold for $1 a share in the summer of 1870, reached $18 in

April 1871, but had sagged back to $6 and $8 late in the year
when Flood began to buy in his quiet way. Lord states that
control of the 11,600 shares in the mine did not cost the Firm
to exceed $50,000, but he cites a low quotation in February.
Shinn suggests:

> They paid, it is said, about $100,000 before they were
> satisfied to announce their control, by which time they
> had about three fourths of the stock.[4]

At the annual meeting of stockholders, held on January 11,
1872, the Firm elected the board of trustees, levied an assessment
of $3 a share, and made preparations to operate the mine. The
trustees elected were Edward Barron, Solomon Heydenfeldt, J. C.
Flood, William S. O'Brien, and B. F. Sherwood. Barron was made
president. Neither Mackay nor Fair went on the board. Fair
was appointed superintendent, but (contrary to his repeated asser-
tions in later life) did not serve as such until nearly a year and
a half later; not until after the bonanza had been discovered.
The first superintendent was T. F. Smith,[5] who was soon succeeded by Captain Sam Curtis, a picturesque and popular old-
timer, under whose supervision the first development work was
done and the discovery of rich ore made.

Their only hope of finding ore was in the Comstock Lode far
below the old barren upper workings, and they first planned to
extend a long drift south and west from the 1100-foot level of
the Ophir shaft.[6]

Fortunately that idea was abandoned, for it would have missed
the bonanza. Then it was decided to sink the vertical Con. Virginia shaft another 500 feet. Bids for sinking were advertised
for,[7] but, luckily, that was not attempted. If the shaft had been
sunk to the 1000-foot level and a crosscut driven westward 600
feet to the Lode, as planned, a probable investment of $200,000
would have been lost. The Lode at that level and below it did
not even contain quartz, much less ore. As fate would have it,
arrangements were then made to prospect the Lode by drifting
north and west into the Con. Virginia from the 1167-foot level
of the Gould & Curry shaft, which stood 1,300 feet south of the

---

[4]The *Story of the Mine*, p. 179 (1896).

[5]*Virginia City Enterprise*, February 16, 1872.

[6]Id., February 1, 1872.

[7]*Daily Territorial Enterprise*, March 31, 1872. Mackay said later that the
machinery was not heavy enough to sink further.

Con. Virginia shaft.[8] Sharon, who was in control of the Gould & Curry, granted permission to do the work through the shaft with the remark: "I'll help those Irishmen lose some of their Hale & Norcross money." All things conspired in their favor. If the drift had been started from the 1000-foot level of the Gould & Curry shaft it would have been almost 100 feet above the top of the bonanza, and the enterprise a failure. Superintendent W. H. Patton, who succeeded Fair in July 1878, said in his report for that year: "The cap rock of the ore body was reached 114 feet above the 1200-foot level."

When Mackay and his associates took control of the mine the stock rose to $30 a share. Then, in the spring of 1872, and before work had been started, the "Crown Point boom" carried it to the ridiculous price of $150 a share, at its highest. That boom was short-lived, but, meanwhile, Flood sold a large amount of the stock at $60 a share or better, which he bought back at high prices when the first low-grade ore was discovered six months later.[9]

Work was begun in the drift about May 1, 1872, and at the same time the west crosscut on the 500-foot level of the Con. Virginia shaft was reopened and development work resumed in the old workings on that level, which proved practically barren.[10]

THE DISCOVERY OF LOW-GRADE ORE

The drift from the Gould & Curry shaft was extended due north until it passed through the north end line of that mine and through the Best & Belcher, a distance of some 800 feet;

---

[8]The shafts of the Ophir, the Con. Virginia, and the Gould & Curry were second-line, vertical shafts, which were sunk 1,200 feet or more east of the croppings in order to intersect the Lode at a depth of from 1,200 to 1,500 feet on its downward and eastward dip.

[9]Testimony of Flood in the Burke (Dewey) case, *San Francisco Bulletin*, December 28, 1880. It appears that he took advantage of the high prices to hedge against possible failure. Control was assured for another year in any event.

[10]It appears that one of the incentives for taking over the Con. Virginia was the hope that some low-grade ore could be found in the old upper workings for the Firm's idle mills, as the ore in the Hale & Norcross was practically exhausted. During the following six months they explored those wide belts of quartz and porphyry, chiefly by means of crosscuts driven every hundred feet across the Lode. The newspapers gave encouraging reports at times, but the work ceased in October 1872. *Gold Hill News*, May 27, June 23, September 6, 1872; *Alta California*, June 24, August 13, October 7, 1872.

In 1875 the Con. Virginia made another unsuccessful effort to find some ore in the old upper workings by reopening the old Latrobe Tunnel. Years later,

then it was turned northwesterly toward the Comstock Lode. On September 12, 1872, at a point 178 feet northwest of the Best & Belcher line, the drift unexpectedly crossed a vein or fissure seven feet wide filled with porphyry, clay, and quartz, and assaying from $7 to $34 a ton.[11] The managers did not know what to think of this vein away out in the hanging wall and leading northeasterly, but low-grade ore had often led to rich ore bodies, and like good miners they turned the drift to follow it. The importance of the discovery in the minds of Mackay and his associates is shown by the fact that Con. Virginia stock rose from $29 on September 13 to $57 on the 19th—the market value of the mine had doubled within a few days. The "Enterprise" of September 19 tells of the discovery and goes on to say:

> Captain Sam Curtis, the present superintendent, is confident of finding good ore in the mine. He says when it is properly opened it will prove to be one of the best mines on the Comstock.

The article says further than the Con. Virginia is grading to put in new and heavier machinery at the shaft, which will be sunk to the 1200-foot level in order to make a connection with the drift coming northward from the Gould & Curry shaft. From this and other statements we learn that Mackay and his associates were developing their new property conservatively. It had not been the intention to go to the heavy expense of installing new machinery in order to sink the shaft deeper unless ore was found below; although Lord states (p. 309) that as soon as the Firm took control, "under the direction of James G. Fair, a large shaft was at once projected and the work of sinking rapidly pushed."[12]

Meanwhile, the drift continued to follow the fissure on its northeasterly course, leading farther and farther away from the Comstock Lode. But the prospects were most encouraging; the low-grade ore gradually increased in size until it was 48 feet in width at the end of the 280-foot northeast drift, and 800 tons of

---

long crosscuts were driven westward to the Lode from the 850-, 1000-, 1100-, and 1300-foot levels of the Con. Virginia and the California, but in each case the Lode proved to be almost barren, the assays hardly averaged over $1 a ton. (Con. Virginia and California reports).

[11]*Daily Territorial Enterprise*, September 19, 1872 and March 20, 1873.

[12]Fair, the best of self-advertisers, caused that statement to be inserted in the Con. Virginia report for 1877. Lord was misled at times by his unfamiliarity with conditions and his reliance upon erroneous statements, although he spent two years on the Comstock in gathering the material for his book.

the rock extracted in drifting averaged $23 a ton, mill returns.[13]
At this point the drift was stopped until the shaft should reach
the 1200-foot level, which was equivalent to the 1167-foot level
of the Gould & Curry. It was during this period of waiting that
the futile effort was made to drive the forked crosscut west to
the Lode.[14]

THE TOP OF THE BONANZA ENCOUNTERED

No sooner had work been resumed in the drift than it entered
a substantial body of good ore. The Gold Hill News of March 1
and the Enterprise of March 2, 1873, tell of the discovery. The
Enterprise details the developments leading up to it and states
that the Con. Virginia began hoisting ore through the Gould &
Curry shaft on March 1. Con. Virginia stock doubled in market
price, $40 to $80, in one day. This was the beginning of the "Big
Bonanza," which was found 1,200 feet vertically below the surface, right in the heart of Virginia City. Contrary to the usual
practice on the Comstock, the public had been kept fully informed
through the newspapers of every development in the mine.[15] By
March 20 the drift had been extended 50 feet, all in ore, some of
it very high grade.[16]

---

[13]*U. S. Mineral Resources* for 1872, p. 110; which says further, "the discovery has every appearance of forming into a very valuable ore body."

[14]Plate 15 of the *Atlas* accompanying *Becker's Geology of the Comstock Lode*
shows at the upper left hand corner the drift from the 1167-foot level of the
Gould & Curry entering Con. Virginia ground. The number of short crosscuts
extending westerly from it toward the Lode indicates that the Firm had not
given up the idea of finding ore there. The ground was saturated with hot
water, the air was bad, and the crosscuts had to be abandoned one after the
other. The longest of them is forked near the end, due to the fact that when
one fork proved impossible they started another. The ground was so loose
that it flowed out on them and the long crosscut had to be bulkheaded to
keep the water and muck out of the drift. The bonanza was found in the most
unstable ground on the Comstock. If the cross-fissure had not been encountered
it would have been impossible to drive westward to the Lode at that time.
R. M. Ballard, foreman, told of that work in the case of Burke v. Flood, et al.
(*San Francisco Bulletin*, December 12, 1880).

[15]"It was the usual policy of those in control of a Comstock mine to keep
secret any improvement until they would secure for themselves a good quantity
of the stock." *History of San Francisco Stock Exchange*, p. 131 (1910).

[16]*Daily Territorial Enterprise*, March 20, 1873.

Mark Twain tells us in his confident way that he knew how the discovery
was made, and proceeds to relate the impossible yarn, fostered by Jimmy Fair,
that the Firm knew in advance that a great deposit of ore existed in the
Con. Virginia because Fair had found it several years earlier when, a day
laborer, groping around an old abandoned tunnel. (Autobiography, Vol. 1, p.
272).

It should be said here, in the interest of historical truth, that the romantic story of the part played by Fair in the discovery of the Con. Virginia bonanza, as told by him to Eliot Lord,[17] is a characteristic fairy tale. The thin seam of ore, which "sometimes narrowed to a film of clay," that Fair claims to have followed "like a bloodhound," was a well-defined vein or fissure, never less than seven feet wide, and bearing low-grade ore at all times. The newspapers and other contemporaneous accounts disprove his story entirely.[18]

Curtis enjoyed but a brief period of glory after the discovery. Through some misunderstanding with Mackay he was relieved on March 24 and Mackay took charge. The Enterprise of March 25, 1873, somewhat indignantly says: "We have heard of no reason for the change. Captain Curtis seems to have been doing all that could have been done by any man for the development of the mine." It appears from a letter in the Con. Virginia files that Fair spent the winter in Oregon with his family. If so, he soon returned and was reelected superintendent. He was a rarely capable mine manager; a grizzly bear of a man physically, of tireless energy and terrific driving power and this most difficult of mines called for all there was in him.[19]

CAPITAL STOCK INCREASED AS BONANZA DEVELOPS

Everything now went forward as fast as capable management could devise and flesh and blood endure. The water and the air in that long lower drift were so hot as to be almost intolerable,[20] but, despite the difficulties, rapid progress was made. Meantime, considerable ore, milling $33 to $35 a ton, was being hoisted through the Gould & Curry shaft.[21] The reduction of ore was begun May 12, 1873 in the Mariposa mill "owned by Mackay and Fair"[22]—for so the mills were known that had been reducing Hale & Norcross ores.

---

[17]Lord's *Comstock Mining and Miners*, p. 310 (1883).

[18]*Daily Territorial Enterprise*, September 19, 1872, and March 20, 1873; also *U. S. Mineral Resources* for 1872, p. 110.

[19]Fair was a slave-driver. He told Lord that "men with families are less vigorous, less energetic, less daring than single men," and had the preference. (Lord's *Comstock Mining and Miners*, p. 381). In later years when it was proposed to raise the salary of the Chief Clerk in San Francisco, he remarked significantly, "A hungry hound hunts the best." (James E. Walsh).

[20]*Daily Territorial Enterprise*, May 22, 1873.

[21]*Gold Hill News*, March 15, 1873.

[22]*Daily Territorial Enterprise*, May 13, 1873.

As the shaft continued downward and the advancing north drift encountered more and better ore the price of the stock gradually increased. In August the Con. Virginia shaft reached the 1200-foot level and a drift was started south to connect with the one coming northward. More important than ore at that time was the question of air in the mine below. Men gasped for breath while they worked in short "spells."[23] At last, toward the end of September, the connection was made and the air at once circulated freely between the Gould & Curry and the Con. Virginia shafts.

The stock, which had been increased from 11,600 to 23,600 shares on May 20, 1872, rose to $100 as the drift continued to advance in ore, and after the top of the bonanza was encountered it reached $240, only to be set back to $48 when the capital was increased from 23,600 to 108,000 shares on October 18, 1873. There was no further change until March 17, 1876, when the capital was increased to 540,000 shares.

It was the policy of the management to keep the price of the stock at a moderate figure—to broaden the market and "give the little fellows a chance," as Flood expressed it—which was accomplished by increasing the number of shares from time to time and declaring stock dividends.[24] The practice of increasing the number of shares in Comstock mines was denounced by the editor of the Mining and Scientific Press on January 16, 1875: "These moves, as everyone knows, are only intended as stock-jobbing operations. 'Giving poor men a chance' is too thin an excuse to be swallowed by anybody." The Bonanza Firm, whose stock transactions were handled by Flood in San Francisco, followed the usual Comstock practice, but played a more open and fairer game. Flood never resorted to tricks and misrepresentation.

Two hundred tons of ore were now being raised daily through the Con. Virginia and the Gould & Curry shafts, and going to the

---

[23]Id., September 4, 1873. The drift at that time was over 1,500 feet from the Gould & Curry shaft and far from straight, which doubled the difficulty of providing ventilation. Such air as they had was forwarded by blowers. Drilling was done by hand. They had no compressed air, which was part of machine-drill equipment. The first of these on the Lode was installed by the Yellow Jacket in the fall of 1872, and was looked upon as an experiment. The Con. Virginia did not install them until the spring of 1874.

[24]"After selling away up in the hundreds of dollars, it was thought best, in order to allow all, rich and poor alike, to trade in these (bonanza) stocks, to still further increase the capital stock." King's *History of San Francisco Stock Exchange*, p. 172 (1910). All of the other Comstock mines had increased the number of their shares from time to time.

Mariposa and Bacon mills. The Occidental mill was being put into commission.[25] A month later, "five of Mackay and Fair's mills" were reducing Con. Virginia ore.[26] The last assessment of $3 a share was levied June 11, 1873, making a total of $277,150.12 under the new management. The payment of dividends of $3 a month on each share was commenced eleven months later.

DAN DEQUILLE'S FIRST REPORT ON THE BONANZA

Soon after a circulation of air was obtained and the workings were in shape for inspection, Dan DeQuille was invited to visit the mine. He was told to take samples of the ore wherever he pleased and to write his own report about what he saw. Dan had been writing encouragingly of the developments during the past year but was quite unprepared for the amount of ore that had been exposed. His first report appeared in the Enterprise of October 29 under the headlines:

"Consolidated Virginia—
A Look Through the Long Forbidden Lower Levels—
The Ore Bodies and Breasts, Winzes and Drifts—
Rich Developments."

The report was both enthusiastic and restrained. "In conclusion we may say that a first-class mine is fast being developed in the Consolidated Virginia, but of course we can see into the ore deposit no further than the openings have been made."[27]

The ore body that Dan inspected was that short segment of the bonanza, 200 feet in length, which was found on the 1200-foot level. The drift from the Gould & Curry shaft had traversed it from the south end, and the drift from the Con. Virginia shaft was turned south and west and entered the ore body at the north end. Crosscuts had been driven across it at intervals, so that the extent of the ore on that level was quite fully disclosed. It was

---

[25]*Daily Territorial Enterprise*, October 22, 1873.

[26]Id., November 24, 1873; *Gold Hill News*, November 24, 1873.

[27]Dan DeQuille looked upon the Con. Virginia bonanza as his baby. As far back as 1867, through the columns of the *Daily Territorial Enterprise*, he had urged the small mines to unite in sinking a shaft in order to explore the Lode at depth, and when the Con. Virginia company was formed for that purpose he lent every encouragement as long as the work continued. Nor did he lose faith when the quartz proved barren on the 500-foot level. Nearly two years later, when Mackay and his associates began operations, Dan was delighted. As the work progressed the venture had his full support, and the great discovery was to him the crowning event of his reportorial career. Alas, for success in prophesy! Dan now began to foretell other bonanzas in the mines all along the Lode—none of which materialized.

from 30 to 40 feet wide and made up of streaks of rich ore separated by belts of mineralized hanging-wall porphyry (diabase). The rich quartz ore, except in the heart of the great ore body, was always accompanied by belts and masses of high mineralized porphyry, most of which was sent to the mills.

The workings had followed the best ore, and Dan's five samples returned high assays; the average being $379.43 to the ton in silver and gold. The following day he returned and took three check samples, which averaged $443.83. The worth of such sampling may be seen in the fact that the average mill return from the ore on that level was $40 a ton. However, it was neither convenient nor desirable to extract the quartz alone, and the inclusion of the porphyry brought down the average. The speculators evidently discounted Dan's enthusiasm, for the price of the stock, which was $51 on the 29th, stood at $51¼ on the 31st.

A miner always wants to know whether his ore is going down, and Fair had sunk several winzes below the 1200-foot level which gave assurance that the ore continued. "God was good to the Irish," laughed James E. Walsh.

The bonanza was formed in two nearly vertical rifts or shear zones in the hanging wall, at and near their intersection, and stood 700 feet east of the Lode itself on the 1200-foot level.[28] The main rift, which extended northerly and southerly and contained nine tenths of the bonanza, broke upward from the Lode at the 1750 level. The rift running southwest and northeast, in which the ore was first found, continued on its course far out in the hanging wall but was unproductive beyond the intersection until some small, rich, disconnected bodies were found years later on the lower levels. The Lode itself was barren to the deepest level. The Con. Virginia shaft, which was sunk about 1,500 feet east of the croppings, reached the Lode on its easterly dip at the vertical depth of 1,500 feet. The C. & C. shaft, located 1,040 feet further east, intersected it at 2,500 feet. In each case the Lode was found to consist of a confused mingling of gouge or clay and porphyry, with no quartz.

---

[28]The position of the ore body in the hanging wall is illustrated on Plate 5 of the Atlas accompanying Becker's *Geology of the Comstock Lode* (1882). He suggested (p. 286) that the rents in the hanging wall in which the bonanza ore bodies were formed were caused by a projecting mass on the diorite foot-wall at the depth of 1000 to 1200 feet, which interrupted the downward slipping of the diabase hanging wall.

Prof. John A. Reid, of the University of Nevada, published a pamphlet on "The Structure and Genesis of the Comstock Lode" (1908), in which he attributed the rifts to unequal movements of the foot and hanging walls.

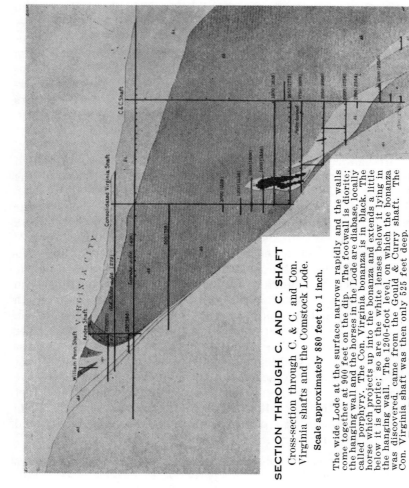

SECTION THROUGH C. AND C. SHAFT

Cross-section through C. & C. and Con. Virginia shafts and the Comstock Lode.

Scale approximately 880 feet to 1 inch.

The wide Lode at the surface narrows rapidly and the walls come together at 900 feet on the dip. The footwall is diorite; the hanging wall and the horses in the Lode are diabase, locally called porphyry. The Con. Virginia bonanza is in black. The horse which projects up into the bonanza and extends a little below it is diorite; so are the white lenses below it lying in the hanging wall. The 1200-foot level, on which the bonanza was discovered, came from the Gould & Curry shaft. The Con. Virginia shaft was then only 525 feet deep.

From Plate 5 of Atlas accompanying Becker's Report.

All of the statements of the production of the mines herein are based on the market value of silver after it was demonetized by the United States on February 12, 1873. The "discount on silver" of the Con. Virginia and California mines alone was $9,826,860 from 1873 to 1897. The oft-repeated statement that the Con. Virginia bonanza brought about the demonetization of silver is an error. The discovery was not made until a few weeks after the law was passed by Congress, and the ore at that time was not of exceptional value. The early fame of the Lode as a producer of silver and the large body of rich ore disclosed in the Crown Point-Belcher bonanza were factors in bringing about this demonetization of silver. Afterward, the wildly exaggerated reports of the quantity of silver in the Con. Virginia bonanza depressed the price of silver in the European markets.

The mine was soon "thrown open to the public," and mine superintendents and other visitors came in throngs to see and appraise the new bonanza.[29] Usually there was a good deal of excitement on the Comstock and in San Francisco when a new discovery was made, but this development was taken very quietly. Perhaps the fact that the ore had been found in such an unexpected place set people wondering how long it would last. The stock hung around $50 for a few months, but nearly doubled in market price before the end of the year, afterward falling back to an average of $80 during the first eight months of 1874.[30]

In the spring of 1872, before commencing operations, the Con. Virginia began to acquire the Kinney, a 50-foot claim which adjoined the shaft on the north, and was thought of slight value. Soon after the discovery of the low-grade ore in the drift the company bought the Central and Central No. 2. At the same time the Firm bought control of the Best & Belcher, adjoining the Con. Virginia on the south, and the Gould & Curry, lying next to the Best & Belcher. Ore bodies on the Comstock had usually pitched southward, and they were forestalling that event, which

---

[29]*Daily Territorial Enterprise*, November 6, 1873.

[30]Speculation in stocks was quieted by the "Panic of 1873," although the West did not begin to suffer until two years later. "In 1873, on September 18, the most extraordinary panic ever witnessed in the United States began. The first three quarters of the year had been prosperous, but on the date mentioned, Jay Cooke & Co. failed, and a financial storm followed which almost destroyed the banking system of the country. * * * The depression that followed lasted until 1877, when more troubles were added by the extensive railroad strike of that year, and there were no signs of recovery until 1878." (*San Francisco*, by John P. Young, vol. 2, p. 504).

failed to materialize, for the ore body was later found to pitch northward.

THE CALIFORNIA ORGANIZED AND MILLS BUILT

Based on the old California and the Kinney, together with the newly acquired Central and Central No. 2, a new company was organized in December 1873 called the California, with 108,000 shares, which was given the north 600 feet; the Con. Virginia retaining the 710 feet adjoining on the south.[31] As it turned out, all of the great bonanza lay in these two mines.

When the California Company was formed each stockholder of the Con. Virginia (which owned $\frac{7}{12}$ths of the new ground) received a dividend of $\frac{7}{12}$ths of a share of the stock of the California, the remaining $\frac{5}{12}$ths being distributed for the remaining portion of the ground so acquired. Now there were two bonanza mines instead of one. Con. Virginia stock sold at $67 a share and California at $37, each with 108,000 shares.

The Con. Virginia and the California at once began to sink a new joint shaft at a point 1,040 feet east of the Con. Virginia shaft in order to facilitate the extraction of the bonanza ore bodies and to tap the Lode on its eastward dip at a vertical depth of 2,500 feet. This shaft, which became known as the "C. & C.," eventually supplanted the Con. Virginia shaft as the basis of operations. The hoisting works and surface plant were the largest and best on the Lode at that time. This was the first of the great third-line shafts.

In December 1873 the Con. Virginia shaft reached the 1300-foot level. Drifts were immediately extended south and east into the ore body, which had increased to 300 feet in length and 50 feet in width. The average value had improved to $50 a ton.

The shaft reached the 1400-foot level in February 1874, and drifts and crosscuts were again sent southward and eastward to the ore bodies, consisting of one similar to that on the 1300 level, with an additional body lying in the north and south zone, but separated from the main body by a large porphyry horse. This ore body was L-shaped, 50 feet wide and 150 feet long. The ore had continued to increase in value with depth and averaged $54 a ton, mill returns, on the 1400 level.

---

[31]There was no need to divide the bonanza between two companies. The purpose was to create a new mine for the stock market. The California was operated through the Con. Virginia shaft until the C. & C. shaft supplanted it as a base of operations.

The ore body on the 1200-, 1300-, and 1400-foot levels lay in the northwest rift or shear zone and took that direction, with the exception of the separate body on the 1400 level, which stood near the intersection of the northwest shear zone with the north and south zone. These levels were not directly below one another, as the ore body constantly pitched downward and northward until it joined the main ore body on the 1500-foot level, which stood almost vertically. In the California the ore barely reached upward to the 1400 level. In the Con. Virginia the ore became narrower and poorer above the 1200 level and terminated just above the 1100 level "on a flat roof."

The Con. Virginia declared its first monthly dividend of $3 a share in May 1874, which was increased to $10 in March 1875. The California did not commence to declare dividends until a year later. The Crown Point-Belcher bonanza began to decline as the Con. Virginia came into production, and was exhausted in 1876.

Cyrus W. Field and family and friends, including Charles Kingsley and other notable English visitors, arrived on the Comstock on May 25, 1874. Several of the party visited the lower levels of the Con. Virginia. A three-decker cage, the first on the Lode, was placed in the Con. Virginia in May. Air compressors and machine drills had been installed earlier.

In June 1874 the Firm began the erection of the Con. Virginia 60-stamp mill, which was located just below the mine. It was an elaborate affair with a capacity of 260 tons a day and cost about $300,000.[32] The mill went into operation in January 1875, but its life was short. It was destroyed in the great fire of October 26, whereupon a similar mill was at once constructed, costing $350,000. That year the Firm also built the California pan mill, with 80 stamps and a capacity of 360 tons in 24 hours, at an expense of "almost $500,000," so Lord reported. Fortunately it escaped the fire, with the exception of the stamp mill, which was then located below the Con. Virginia dump. Although called the Con. Virginia and California mills, neither company ever owned a mill of its own; the ore from both mines was reduced under contract at ruling prices in these mills owned by the Firm. The

---

[32]The mill employed the Washoe process as perfected years before, and differed only in being more perfect mechanically. There were 32 amalgamating pans, each 5½ feet in diameter and holding 3,000 pounds of pulp, and below them 16 settling pans and 8 agitators. W. H. Patton designed and supervised the erection of the mill, which is described in detail in Dan DeQuille's *Big Bonanza*, pp. 336–345 (1876).

local and San Francisco newspapers speak of that as a matter of course: "Mackay, Fair, Flood, and O'Brien have just incorporated their mills under the name of Pacific Mill and Mining Company. Trustees: O'Brien, Flood, Fair, Mackay, and Barron."[33]

In July the Ophir discovered a small body of ore on its 1465-foot level near the north line of the California. The ore was of

CALIFORNIA PAN MILL

Built by the Bonanza Firm in 1875. The crushing plant was located below the C. & C. Shaft and the pulp flumed to the mill. The C. & C. hoisting works shows at the left center; beyond it is the Con. Virginia, with four stacks. The Con. Virginia mill is in the center of the picture, with the Ophir hoisting works and dump at the extreme right showing faintly.

the same character as that in the California and stood in a similar shattered zone, although running more easterly and westerly, which led to the assumption that the bonanza extended through the California and into the Ophir.

---

[33]*Daily Territorial Enterprise*, August 14, 1874. This time Dan wrote the name correctly.

Presently the mining experts began to make enthusiastic reports. The stock market, however, failed to respond, which brought complaints from the newspapers: "Stocks appear to grow weaker as developments in the mines multiply and the faith of our mining men in the value of these developments strengthens."[34]

---

[34]Id., August 1, 1874.

All of the leading mines were sinking their shafts as rapidly as possible. The Savage and Hale & Norcross had reached 2,200 feet and were soon to be baffled by floods of water. The other main shafts averaged 1,700 feet.

CHAPTER XVIII

Sharon Starts the Boom—A Wild Market—The Consolidated
Virginia Boom—The Chronicle Boosts the Bonanza—The
Market Reaches the Top—Extravagant Forecasts.

It takes a leader to start a boom. Con. Virginia was expected
to do that but the members of the Firm were content to watch
their stocks increase in value and collect their regular monthly
dividend of $3 a share.[1]

It fell to William Sharon, an avowed candidate for the United
States Senate, to start the boom. He had lived in San Francisco
after Jones defeated him for the Senate in 1872, meanwhile grow-
ing richer month by month from the Belcher and the Union
Milling Company, and nursing his senatorial ambition. When
Stewart announced that he would not seek reelection Sharon
again entered the lists, this time with full determination to win,
whatever the cost. He first bought the "Enterprise" from editor
Goodman who had flayed him in 1872, thereby acquiring a champion, then set out to obtain control of the Ophir in order to use
the stock to further his campaign.[2] Besides, the mine itself was
promising to develop an extension of the Con. Virginia bonanza,
which would add to Sharon's prestige and furnish ore for his mills.

Unfortunately for his plans the control of Ophir was in the
hands of E. J. "Lucky" Baldwin, one of the shrewdest men on the
Coast, who had been content to let the stock ride along quietly.
Sharon found him a hard trader. As the annual meeting of
stockholders was to be held on the 13th of December, it became
necessary to acquire over one half of the 100,800 shares before
that day in order to elect the new board of trustees.

Sharon began to buy quietly. On August 11, 1874, Ophir stood
at $20 a share, Con. Virginia at $80, and California at $40. A
month later Ophir reached $52, while the two bonanza stocks had
advanced but a few dollars. The sharp rise began toward the end
of October, after Sharon bought James R. Keene's block of Ophir
stock and employed him to manipulate the market. "Jim" Keene

---

[1] The *Story of the Mine*, p. 182 (1896).

[2] Davis' *History of Nevada*, Vol. 1, p. 421; Charles De Long Letters in California State Library.

Sutro was Sharon's only active opponent on the Republican side. Sharon
took him seriously, but Sutro again failed to get a single vote in the Legislature.

was a genius but he could not pry Baldwin loose. Sharon not only began to buy but to sell "short" at the same time. He was willing to pay a high price for the stock and to agree to sell it back within 90 days at a much lower figure. Meantime the Ophir meeting would be over and the vote for Senator as well. King tells of Sharon's short sales on pages 67 and 171 of his "History of the San Francisco Stock Exchange," of which he was a member.

Throughout November the market was in a ferment. Ophir reached $100 a share, Con. Virginia $160, California $90, and all of the other stocks rose with them. A speculator could buy any stock and sell it within a few days at an advance, which creates a boom in any market. The wild market that followed was caused not only by Sharon's manipulations, but by the amazing developments on the 1500- and 1550-foot levels of the Con. Virginia.

Baldwin withstood all of Keene's blandishments, and Sharon was forced to pay his price for his stock, $135 a share for 20,000 shares.[3] Sharon took over control at the Ophir election, but kept boosting the stock until he was elected Senator on January 12, 1875.[4] Meantime the speculators had gone mad: Ophir sold for $315 a share on January 7, Con. Virginia for $710, and California for $780. The inevitable panic started on January 8, and the bottom fell out of the market.

It was charged that Sharon had "unloaded" at high prices and then "shorted" the stock, thereby recouping all of his expenditures.[5]

Ralston, a silent partner, was left high and dry. That is confirmed by Sharon's biography (written six years after his death), which, by one of those not uncommon distortions of history, says it was Ralston who battled for the control of Ophir in a desperate effort to retrieve his failing fortunes, and that "it ruined him and the bank."[6] Ralston was already a ruined man; the Ophir was merely an incident.

---

[3]*Virginia Evening Chronicle*, December 12, 1874; Reuben Lloyd's ms. in Bancroft Library dated 1886.

[4]Sharon's wealth and the high-handed methods pursued in acquiring it proved the chief stumbling block in his campaign. One night at a rally in the Opera House he closed an earnest defense with the statement: "You know I can't take my money with me." Whereupon a voice from the gallery responded "If you did it would burn."

[5]Davis' *History of Nevada*, Vol. 1, p. 421; Letters of Charles De Long to his wife, in California State Library.

[6]Bancroft's *Chronicles of the Builders*, vol. 4, pp. 23–65.

AMAZING DEVELOPMENTS AND A WILD MARKET

But to return to the month of October, when the developments on the 1500-foot level of the Con. Virginia and the California began to amaze the most optimistic. The mines now began to be crowded with superintendents, engineers, stock speculators, and others, mostly from San Francisco, all eager to examine the new bonanza. So great was the throng that work was constantly impeded throughout the fall. Dan DeQuille says: "Never before in the history of the Comstock was the public so fairly treated."[7]

The members of the Firm not only wanted to deal generously, but they were justly proud of their new mine, although with little display of personal pride in their sudden elevation to the topmost round of the mining world. "Mr. Mackay is a very unpretentious man for a millionaire," commented the San Francisco Chronicle. The Enterprise noted the following visitors on September 22:

> The following boss manipulators of stocks have been viewing the situation along the Comstock for a day or two past: William Lent, Robert Sherwood, George Hearst, J. W. Pierson, R. F. Morrow, D. L. McDonald, M. D. Townsend, and Charley Forman.

When Con. Virginia was selling around $100 a share, Robert Sherwood, who had 1,000 shares, told his friend Flood one morning that he was getting tired because the stock did not move. "What are you growling about?" replied Flood. "If you are tired of that stock, I'll take it off your hands at $100." A year later, after the Firm had built the fine Nevada Block on the corner of Pine and Montgomery Streets, Flood remarked to Sherwood one day: "We built that block from the profits on that 1,000 shares." Sherwood waited four years to even the score. At the height of the "Sierra Nevada Deal," in September 1878, Fair agreed to buy his 5,000 shares of Union at $200 a share and Flood turned over $1,000,000 in government bonds to pay for it. Sherwood invested the money in San Francisco real estate, and later remarked to Flood: "I built the Union Block with the profits on that 5,000 shares."[8]

A glowing account of the bonanza appears in the "Enterprise" of October 25, 1874. It makes the extravagant statement that the great ore body is now known to extend from the Gould & Curry through the Con. Virginia, California, and Ophir to the Union Consolidated, and ventures the roseate suggestion:

---

[7] Dan DeQuille's *Big Bonanza*, p. 473.
[8] King's *History of San Francisco Stock Exchange*, p. 155 (1910).

At the further depth of four or five hundred feet in their mine the Consolidated Virginia may reach a point where their ore will lack but little of being solid silver. Specimens brought from the drift running into the Consolidated from the 1500-foot level of the Gould & Curry are even now almost masses of silver.

This long drift from the 1500-foot level of the Gould & Curry shaft was driven to explore the bonanza along its length in the Con. Virginia and the California. The "Gold Hill News," whose reporter visited the bonanza on October 23, said that much of the ore in the drift is so rich that it is sacked and hauled to the mill where it requires 500 pounds of quicksilver to the pan as against the usual 200 pounds, and added: "The future prospects of Con. Virginia are almost beyond estimate."

Now developments began to be startling. A crosscut on the 1500-foot level of the Con. Virginia which crossed the ore body diagonally where it turned northward at the intersection of the two rifts or shear zones, was in rich ore for 300 feet, and when the drift from the Gould & Curry passed through a body of rich ore 400 feet in length in the Con. Virginia and then through a similar body in the California that was even richer, the excitement among mining men and the general public was unbounded. On December 3 Con. Virginia sold at $196, California at $160, and Ophir at $118, which led Dan DeQuille to suggest:

> There are those who predict that California will go up to $1,000—mining men at that. * * * All will now await with almost breathless interest the result of the crosscuts now being advanced across the vein well out in the center of California ground.[9]

The Consolidated Virginia had been obliged to cease admitting visitors during the next week or two. "The miners on the day shift on the 1500-foot level have not been able to work more than half of the time."[10]

The heart of the bonanza was disclosed on the 1500- and 1550-foot levels. There the ore occurred in two large bodies, one in the Con. Virginia that extended into the California, and another in the California. These two bodies were separated by a porphyry horse, but as it was sufficiently mineralized to pay to mine and mill the stopes on those levels eventually joined. Becker's stope map shows that a continuous body of ore was mined on both levels

---

[9]*Daily Territorial Enterprise*, December 8, 1874.
[10]Id., December 9, 1874.

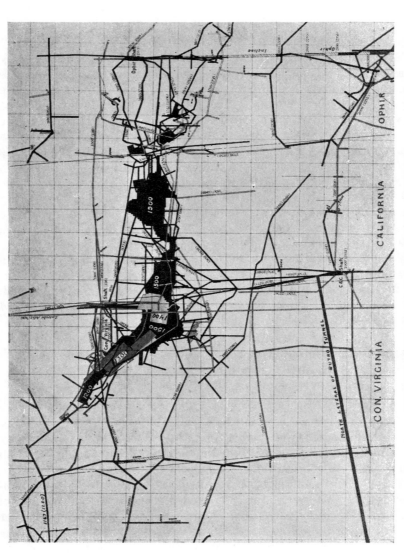

HORIZONTAL PLAT OF BONANZA STOPES

(Description, opposite page)

for a length of fully 900 feet. Its greatest width, 200 feet, was at the intersection of the shear zones. Although this great deposit was enriched by narrow veins of ore that assayed hundreds and sometimes thousands of dollars to the ton, the mill returns in 1876, when the richest ore was mined, averaged only $105 from Con. Virginia and $98 from the California. The average recovery of both mines from 1873 to 1882 was $75 a ton, which Fair explained in his reports for 1875 and 1876:

> In extracting the ore, instead of trying to show an extraordinary yield by the working of the higher grade alone, it has been my constant purpose to so unite the higher and lower grades as to make the average yield to conform to an average standard value.

Becker said of the ore in the bonanza:

> It was composed of crushed quartz, including fragments of country rock, and carried a few hard, narrow, vein-like seams of very rich black ores, consisting of stephanite and similar minerals, while nearly the whole mass of "sugar quartz" was impregnated to a moderate extent with argentite and gold, the latter probably in a free state. The immense volume of these soft ores more than compensated for their moderate tenor, and much the greater part of the entire yield of the bonanza was derived from them. They carried a moderate amount of pyrite. A great part of the space stoped out consisted of fragments of country rock, impregnated, however, with ore, and assaying well.[11]

HORIZONTAL PLAT OF STOPES

The plat on the opposite page is a composite horizontal section, showing the Con. Virginia and California bonanza from the 1200- to the 1500-foot levels, and the three small Ophir ore bodies adjoining, together with the location of the three shafts and the principal mine-workings (drifts, crosscuts, and winzes) from the 1200- to the 2200-foot levels. In other words, we are looking down from an assumed flat surface on the 1200-foot level to other assumed flat surfaces on the levels below. Each square represents 100 feet.

The 1200 drift from the Gould & Curry, in which the bonanza was discovered, appears in the upper left corner. The stopes on

---

[11]*Geology of the Comstock Lode*, pp. 269, 270 (1882).

the 1200, 1300, and 1400 levels were in the rift extending southwest and northeast.

The Con. Virginia ore body on the 1400-foot level is nearly covered by that on the 1300, with the exception of the large angular, disconnected body on that level, which merged into the stopes on the 1500-foot level.

The bonanza stood nearly vertically below the 1500-foot level, whose immense stopes cover those on the 1550, 1600, 1650, and and 1750.

The stopes on the 1550-foot level were substantially of the same extent as on the 1500, but the height to the 1500 level was only 42 feet. On the 1600 (only 58 feet below the 1550), the ore body suddenly contracted in size and diminished in value. In the Con. Virginia it was only 200 feet in length and 65 feet in width, with another of like dimensions in the California, the two bodies being separated by a porphyry "horse."

On the 1650 level the ore body was still smaller and of lower grade. In the Con. Virginia it was 200 feet in length and 90 feet in width, but filled from end to end by a "horse" of porphyry 30 feet thick. The ore body in the California on that level was of substantially the same dimensions.

The bonanza practically terminated on the 1650-foot level. Below that it grew thinner and poorer, and all that was left of the good ore on the 1750 level was a body 40 feet long and 28 feet wide. Some narrow sheets of fair ore were found down to the 2150 level.

The Con. Virginia and C. & C. shafts were vertical, and encountered the barren Comstock Lode on its easterly dip at the depth of 1500 and 2500 feet, respectively.

The bonanza was formed at the intersection of two shear zones or rifts in the hanging wall, one extending southwesterly and northeasterly, and the other northerly and southerly. The solutions that enriched the bonanza appear to have ascended the northeast rift, which carried small, disconnected bodies of rich ore, pitching northward, from the 1800-foot level of the Con. Virginia to the 2500-foot level of the Mexican. These disconnected bodies were discovered from 1901 to 1911 when the C. & C. shaft was pumped out to the 2500-foot level. The northeast rift, however, did not contain ore from the 1800 level to the 1500.

(From Atlas, Sheet 15, accompanying Becker's report, which shows only the mine workings to the year 1880.)

THE CHRONICLE BOOSTS THE BONANZA

The "San Francisco Chronicle," then the leading newspaper of the Pacific Coast and boasting of the largest circulation, led all the rest in extolling the riches of the bonanza and the virtues of its managers. Because of the leading part that newspaper played in boosting the market to such tottering heights, and because in later years it hounded the Bonanza Firm with almost unbelievable violence, its laudatory articles and editorials are given prominence in the following pages. Its denunciations will be set forth in connection with the suits brought by Squire P. Dewey against Mackay and his associates, which were inspired and supported by the Chronicle. The change of heart of the Chronicle was ascribed by Flood and his friends, and by the other newspapers,[12] to the unsuccessful attempt of Charles de Young,[13] chief owner and manager of the Chronicle, to induce the Bonanza Firm to make good his loss on Con. Virginia stock, which he bought in December 1874 after an inspection of the mine accompanied by an expert.

Although the boom was started by Sharon, assisted by those in control of unprofitable mines, its immense inflation was made possible by the most important discovery ever made on the Comstock, which unbalanced men's judgments and aroused the cupidity of the gambling public. At that time there were but three Comstock mines paying dividends—the Con. Virginia, the Crown Point, and the Belcher. The Crown Point paid its last in 1875, the Belcher in 1876.

The Chronicle of December 9, 1874, printed on its front page a moderately enthusiastic description of the bonanza, written by Dan DeQuille, and accompanied by a rough vertical section of the ore body from the 1200 to 1500 levels.

On December 20, 1874, the Chronicle carried a column entitled

[12]*San Francisco Call*, April 29, 1877; *Daily Stock Exchange*, April 1, 1878; *Daily Territorial Enterprise*, April 5, 1878; *Alta California*, January 13, 1877.

[13]The *San Francisco Chronicle* was published by "Charles de Young & Co., Proprietors," until after his death at the hands of young Kalloch on April 23, 1880. Kalloch was acquitted. Eight months earlier, on August 25, 1879, de Young had shot and dangerously wounded the father, Rev. I. S. Kalloch, then a candidate for mayor, after an exchange of scandalous personalities in the newspaper and from the rostrum, and the detraction had continued. (Hittell's *History of California*, Vol. 4, pp. 656, 657 and current newspapers.)

The names of Chas. de Young and M. H. de Young appeared above the firm name after September 21, 1874, and, beginning April 11, 1883, the name of M. H. de Young appeared as "Proprietor."

"Fortune's Flood Tide": "Day after day advices from the Big Bonanza show increased value by more extensive developments, proving riches before which the treasures of Aladdin's Palace and Monte Cristo's fabulous island pale." The article gave the names of sixty men (other than Mackay, Fair, Flood, and O'Brien) who are reputed to have made fortunes in Con. Virginia, among them R. N. "Bob" Graves, who had 8,000 shares of Con. Virginia, $2,000,000, and General Tom Williams over $2,000,000.

The Chronicle editorially lauds the generosity of the members of the Bonanza Firm in sharing their good fortune with old-time friends:

> One of the most agreeable features of the discovery of the new bonanza is that a great number of poor people have reaped substantial benefits from it. This is the more noticeable because it is new in the history of stock operations. * * * When the Consolidated Virginia bonanza was discovered instead of the managers seeking to gobble it all up for themselves, they not only made no secret of the discovery, but actually advised their poor friends to purchase. We hear of scores of poor men and women, who, getting information directly from the managers and leading owners of the mines, have made fortunes by its advance in price. This is a bright spot of good honest charity in a desert of fraud and subtle iniquity.[14]

General Thomas J. Williams and David Bixler, who had been associated with Mackay in the luckless Bullion, were among the first to acquire large fortunes in Con. Virginia. About 1865, when mines had little value, they brought suit against Central No. 2 for attorney's fees, amounting to $1,200, for services rendered several years earlier. The company did not defend and they bought in the mine for the amount of the judgment at the resulting sheriff's sale. It is said that their fellow lawyers used to josh them about their mine.

After the Con. Virginia bonanza was discovered and that company began to acquire adjoining properties, Williams and Bixler exchanged their interest in Central No. 2 for a few hundred shares in Con. Virginia and California. They assisted also in purchasing interests in the Kinney and other claims for the

---

[14] *San Francisco Chronicle*, January 10, 1875.

King confirms this on page 172 of his *History of the San Francisco Stock Exchange*.

company, for which they were paid in stock. Owing to successive increases in the capital stock of Con. Virginia and to the organization of the California, their holdings in 1874 had increased to 6,500 shares of California and 3,000 of Con. Virginia,[15] which they sold at nearly the height of the boom "for upwards of $3,000,000," according to Marye,[16] although the amount was generally stated as $4,000,000. Again, wisely they invested the money in California real estate. They were the only Comstock lawyers to acquire and retain substantial fortunes. The others that made money lost it in the stock market.

THE MARKET REACHES THE TOP—EXTRAVAGANT FORECASTS

The "Chronicle" of December 23, 1874, says that the "Call" and the "Bulletin" have been criticizing and sneering at its alleged sensational items and editorials boosting the bonanza, and that the "Bulletin" charges it with a "violation of journalistic propriety" in so attempting to inflate the stock market. Meantime, those two newspapers had been warning the public that the market was greatly inflated, and to beware. The "Chronicle" justifies itself in the fact that stocks have doubled in value since December 9, when it published Dan DeQuille's first description of the bonanza. It suggests also that "Sharon's modest estimate of $300,000,000 * * * impeaches him as an unskillful cipherer."

The "mining experts," who were all interested in the market, lauded the bonanza to the skies, but the most preposterous forecast was made by Philipp Deidesheimer, the noted Comstock engineer and mine superintendent. He gave a sensational interview on the bonanza mines to the "San Francisco Post" on December 21, 1874, reprinted in the "Chronicle" of December 23, 1874, in which he asserted that the bonanza in the Con. Virginia, the California, and the Ophir would yield $1,500,000,000. "Nothing like these mines has ever been seen or heard or dreamed of before," he said.[17]

---

[15]*Virginia Evening Chronicle*, December 5, 1874.

[16]Marye's *From '49 to '83*, pp. 107, 179 (1923).

[17]Deidesheimer backed his judgment by buying bonanza stocks on margin account, which brought him to bankruptcy a few months later, with liabilities amounting to $534,600, assets none. His genial nature and optimistic temperament brought him employment in many mines. Sharon made him superintendent of the Ophir early in 1875. After twenty years on the Comstock as superintendent of various mines, Deidesheimer went to Sierra City, California, where he became one of the five fortunate owners of the Young America mine, which paid large and regular dividends for five years. As often happens, that large and rich ore body terminated abruptly, and terminated also the lavish spending career of the owners. Deidesheimer never had another streak of luck, although he lived to a great age.

Superintendent Fair's report of Con. Virginia operations during the year 1874, which tells of the great widths of high-grade ore developed on the 1500- and 1550-foot levels, "is exultant and leaves room for the wildest imaginings," but is not actually misleading as were his reports for 1875 and 1876, which were used with damaging effect by the critics of the Bonanza Firm. The 1874 report was published on December 31, almost at the height of the boom, and drove the speculators to still greater frenzy.

On January 5, 1875, the "Chronicle" published another front-page article by Dan DeQuille, with sensational headlines, in which he describes the ore bodies extending through the Con. Virginia and the California and into the Ophir, the 1550 level being illustrated by a rough and misleading sketch. After a glorified description of the bonanza developed from the 1400 to the 1550 levels in the Con. Virginia, Dan estimated the ore in sight in that mine between those levels at $116,748,000, assuming that he was dealing with a body of ore 220 feet wide and 300 feet in depth, of an average value of $100 a ton, which he suggests is only half of its real value.

Dan's estimate was a modest one at the time, if, as he and others assumed, he was dealing with a known ore body of regular size and shape. But that ore had been formed in an irregular rent in the hanging wall and varied in width and in value almost from foot to foot. It was most fully exposed on the 1500-foot level of the Con. Virginia in three crosscuts which had been driven across the ore body, each 100 feet apart: Crosscut No. 1, which was 20 feet south of the California line, showed 145 feet of ore assaying $900 a ton; crosscut No. 2, sampled $80 a ton for 70 feet and $175 for 145 feet; crosscut No. 3, driven partly on a curve, showed 150 feet of $200 ore.[18] It did not then appear that the crosscuts had missed some leaner ore, including a considerable amount of porphyry, and when that ore came to be mined it was found that the crosscuts indicated more than twice as much ore as was extracted between the 1400- and 1500-foot levels. That ore milled $95 a ton, after losing 20 percent of the values in the tailings.

The following editorial from the "Chronicle" of January 4, 1875, illustrates the exuberance of its enthusiasm. Under the

---

[18]The length of the three crosscuts and the assay value of the ore in each are shown on a plat on page 198 of Raymond's *U. S. Mineral Resources* for 1874, published in 1875. Raymond, however, after a cool analysis, declined to allow more than $60,000,000 "in sight" in both the Con. Virginia and the California in March 1875, which proved to be a remarkably close estimate.

caption "Millionaires of San Francisco," the editor quotes the following paragraph from the "Chicago Inter-Ocean":

> No city upon this continent can show more men of solid wealth than San Francisco. Mines of fabulous possibilities pour their dividends into the pockets of the Licks, the Sharons, and the Haywards. Many of her citizens could sell out at a month's notice for $5,000,000 each. Palaces have risen from silver bricks, and the proudest buildings in the City owe their origin to ores and bullion.

Following which the editor of the "Chronicle" proudly comments:

> This was true enough three months ago, but the Inter-Ocean is one of those old fogy journals who do not keep pace with the times. Lick, Latham, Sharon, and Hayward are all poor men. Worth $5,000,000? Well, yes, they may be worth that paltry sum. So are Reese, Mills, Baldwin, Lux, Miller, Jones, Ralston, and Stanford. These are only our well-to-do citizens, men of comfortable incomes—our middle class. Our rich men the Inter-Ocean has not named. They are Mackay, Flood, O'Brien, and Fair. Twenty or thirty millions each is but a moderate estimate of their wealth. Mackay is worth from sixty to a hundred millions.
>
> They have not heard of our new bonanza in Chicago— a lump of silver ore as big as their Grand Pacific Hotel, worth from $100 to $20,000 per ton, so far as ascertained, while its depth, breadth, and thickness have not yet been reached.

Nothing like this ore body had ever been seen in the world, so most of them thought.[19] It had continued to increase in size and richness from the 1200- to the 1550-foot level, and what might be developed below stirred the wildest imaginings.[20] It was beyond thought that this great and rich bonanza would suddenly contract in size and decrease in value only 58 feet below the

---

[19] Although there had been greater and richer deposits of silver ore at Potosi and in Mexico.

[20] The richest ore in the mine lay between the 1500 and 1550 levels. The richest spot was a small body on the 1500 level, close to the south line of the California. Dan DeQuille's *Big Bonanza*, p. 485 (1876), says that a chamber ten feet square was opened there whose "walls were a solid mass of black sulphuret ore flecked with native silver, while the roof was filled with stephanite, or silver in the form of crystals."

1550-foot level, on the 1600 level, so-called, and practically terminate on the 1650 level.

The stock brokers were almost exhausted by the rush of business, and on December 24 the Exchange declared a recess until January 2. But there was no holding back the flood. Stocks continued to mount in street trading, and on January 7, 1875, the market value of 31 of the leading mines was $262,669,940.[21] A tabulation of the 65 others would increase the amount to $300,000,000. The market value of Con. Virginia was $76,680,000, of California $85,380,000 and of Ophir $31,748,000. What that sum meant at the time may be illustrated by the fact that the total assessed value of the real estate in San Francisco was only $190,000,000.[22]

It appears that the Firm had caught the contagion and was buying at the top of the market. "Con" Sullivan, a successful speculator, had bought 100 shares of Con. Virginia at $8 a share in December 1871, and retained it, together with the increases, until the latter part of December 1874, when Flood offered $680,000 for his shares in Con. Virginia alone. Sullivan asked for an hour to consider and then accepted.[23]

Now the boom was at its height. The people of the Pacific Coast had gone mad under the suggestion of a constantly rising market and the wild stories about the ore in the bonanza mines and the prospects of additional great ore bodies in all of the other mines.

Thousands of men and women of all classes had rushed into the market to purchase stocks in Comstock mines.[24] Few of them could afford high-priced stocks and so bought the cheaper ones which were being manipulated by the "insiders."

A characteristic example of the prevailing delirium is given by

[21]Lord's *Comstock Mining and Miners*, p. 409 (1883). On that page Lord gives the market value of each of those mines.
[22]*San Francisco Bulletin*, January 27, 1875.
[23]King's *History of San Francisco Stock Exchange*, p. 245 (1910). Each of the original 11,600 shares of Con. Virginia had increased to 9.31 shares, through successive increases in the capital stock, first to 23,600 and then to 108,000 shares. In addition, each of the latter shares received 7/12th of a share of California when that company was formed in December 1873. In round numbers, a person that had purchased 100 shares of Con. Virginia at $1 in 1870, and retained his stock, would now have 931 shares of Con. Virginia and 543 shares of California, in all 1,474 shares, which could have been sold, at the peak of the market, in January 1875, for the sum of $1,089,980. Meantime, the dividends had more than repaid the assessments. An increase unmatched in the history of mining.
[24]Thompson & West *History of Nevada*, p. 619 (1881).

Rollin M. Daggett in the "San Francisco Call" of September 10, 1893: Dr. Bronson, a noted Comstock surgeon, and an inveterate stock gambler, had made $500,000 in the Crown Point boom in the spring of 1872; then got caught in a declining market and had to go through bankruptcy. In January 1875 he had 1,000 shares of Con. Virginia, bought on margin account, and could "clean up" over half a million dollars. "Sell that stock for the beggarly price of $700 a share!" he exclaimed. "I would rather kindle a fire with it, and I will before I'll sell it for less than $3,000." Again the market broke on him, and again the Doctor went into liquidation.

It was reported that Coll Deane (a prominent broker), bought of James R. Keene 1,000 shares of Consolidated Virginia at $800 a share (buyer ninety), which means that the purchaser was bound to pay for the stock at that price within ninety days, and that Deane took a heavy loss. "It was thought and believed at the time that the stock would reach $3,000 a share."[25]

---

[25]King's *History of San Francisco Stock Exchange*, p. 153 (1910).

The only stocks that had not materially advanced in price were the Crown Point and the Belcher, and they were the only mines on the Lode that were paying dividends with the exception of the Con. Virginia. Evidently they had not been manipulated.

CHAPTER XIX

The Fateful Year of 1875—The Market Crashes in January—
Mackay Thought the Panic Was Temporary—The Chronicle
Attacks "The Bank Crowd"— Partial Recovery in the
Spring—The New York Tribune Correspondent Visits the
Comstock—Prof. Rogers and the Director of the U. S. Mint
Overestimate the Bonanza—The Firm Organizes the Nevada
Bank—The Bank of California Suspends in August 1875—
Ralston's Death Stuns the Coast—Virginia City Destroyed
by Fire October 26, 1875—Fair's Misleading Reports.

Suddenly the speculators came to their senses. Practically all
stocks had been bought on margin account and there was not
enough money in the West to finance more than a fraction of
the purchases.[1]

Even if the Comstock mines had been worth the price quoted,
it would have been impossible to maintain such a ratio between
the amount of available capital and the stock valuation, but many
of the mines were not worth a dollar intrinsically and all were
overvalued.[2]

The market had reached the top and could go no further. First
it hesitated. The street speculators began to sell and prices
dropped. Then the public took alarm, "like a flock of pigeons,"
and began to throw their stocks on the market. With every sharp
decline more margin accounts were uncovered and bankers, brokers, and money lenders dumped the shares held as security in
order to protect themselves. The speculators in San Francisco
were panic-stricken.

On January 8, 1875, the "San Francisco Bulletin" reported:
Ten leading Comstock mines depreciated $17,814,800 in value in
the last twenty-four hours; California lost $70 a share, Con. Vir-
ginia $30, Ophir $40.

---

[1]"The prices of stocks listed on the Board in January 1875 footed up to a
grand total of $350,000,000. (This included other mines as well as Comstocks),
and we were compelled to rely solely on the Pacific Coast for this amount in
gold." All business transactions in California and Nevada were on a gold
basis, by common consent, from 1862 to 1879, when the rest of the United
States was on a depreciated currency basis. King's *History of San Francisco
Stock Exchange*, pp. 22, 46, 283–289 (1910).

[2]Lord's *Comstock Mining and Miners*, p. 316 (1883).

The editor of the "Chronicle" on January 9 maintains his faith that the bonanza mines are not overvalued, but warns of the dangers of margin acounts.

Flood was induced to give an interview to a "Chronicle" reporter on January 11, in which he said:

> We have got a mine which proves vast and rich beyond our most sanguine expectations. What its extent is, exactly, we cannot tell any more than you can. Everyone who desires can go into the mine and see what it contains for himself, and not only that but can take out specimens of ore and have them assayed wherever he pleases. We are going to run the mine for the benefit of all the stockholders; we have the majority of the stock, and it is to our own interest to make it pay as much as possible.

When asked by the reporter about the price charged by the Bonanza Firm for milling Con. Virginia ores, Flood went into detail to show that the present charge of $13 a ton was fair. The ore, he says, is very expensive to mill; it requires three times as much quicksilver as most other Comstock ores, and the cost of milling is about $10 a ton. He stated also that they were working the ore up to $77\frac{1}{2}$ percent of its assay value.[3]

A few years later, when the "Chronicle" was hounding the Bonanza Firm, it affected surprise that the Con. Virginia was not milling its own ores.

The "Chronicle" later jibed at Flood, calling it "vain glory" for exhibiting at his club, about this time, one certificate for 60,000 shares of Con. Virginia to show that he and his associates had not been selling their stock.

The decline in the prices of all Comstocks in thirty days was appalling. Some wise bears, notably James R. Keene, had "shorted" the market and reaped a rich harvest. Keene is said to have been worth $5,000,000 when he left for New York in the latter part of 1876.[4]

---

[3] J. B. Hereford, superintendent of the Bonanza Firm's mills, stated in 1875, when they were milling rich ore, that the loss in quicksilver was from two to three pounds to the ton. Owing to the largely increased demand, the price that year was $1.20 a pound. During 1874, it was $1.40 a pound. Prior to 1873 it had averaged about 70 cents a pound.

The milling charge was reduced as quicksilver became cheaper, and the ore lower in grade, to $12 a ton in October 1876, to $11 in February 1877, in April 1877 to $10, and in January 1878 to $9.

[4] King's *History of San Francisco Stock Exchange*, p. 126 (1910).

MACKAY THOUGHT THE PANIC WAS TEMPORARY

An interview with Mackay was printed in the "Virginia Evening Chronicle" of January 28, 1875, in which he said that in his opinion the decline was only temporary and that the market would soon rally.

The reporter then asked L. P. Drexler, a prominent stock broker, the cause of the crash:

"Well, I think it is no more than the natural decline, after a rise so sudden and unexpected."

"Do you not believe that the mines are worth what they have been selling at?"

"Indeed I do not, nor ever have. I think that they went up on expectations which could never be realized."[5]

Mackay had declined to make an estimate of the worth of the mines during the boom, but on February 7, after the collapse, in order to disavow any hand in the market and to reassure the public, he gave to Dan DeQuille the last long newspaper interview of his career, from which it appears that he was almost as optimistic as the experts over the future of the bonanza mines:

> In a conversation with Mr. Mackay a day or two since in regard to the bonanza, he said that it was indisputably the greatest mass of silver ore ever discovered in any place in the world. "Ten years from now," said he, "people will all know and admit this." In regard to the Consolidated Virginia, he said: "Some think that the stock has already sold for more than it is worth. The truth is that it has never yet sold for one half of its value, but all this will be seen in good time. People will see it after a while." Speaking of the great crash in stocks, Mr. Mackay said: "It is no affair of mine. I am not speculating in stocks. My business is mining, legitimate mining. I see that my men do their work properly in the mines and that all goes on as it should in the mills. I make my money here out of the ore. * * *" During one short conversation Mr. Mackay repeatedly said: "My business is square, legitimate mining. I make my money here from the mines. Here and in San Francisco persons are constantly coming to me and asking me, or

---

[5]King (p. 199) says that "Drexler made a large fortune by selling bonanza stocks at high prices." It is said that he was on the short side of the market in the "1886 Deal" (when James L. Flood, son of James C., put up a job on the shorts) and was compelled to liquidate.

writing to me, to ask: 'What shall I buy?' In San Francisco they regularly besiege me. I said to all that came: 'Go and put your money in a savings bank.' "[6]

Neither Mackay nor Flood was competent to pass judgment on the mine or on the market. They were dazzled by the vast riches exposed and thrown off balance by the enthusiasm of their friends. They had much to learn about the ways of the stock market, and learned it bitterly. Both Lord and Shinn comment on Mackay's steadfast refusal to make an estimate of the ore in sight, but the effect of his interviews in the "Chronicle" and the "Enterprise" was the same as if he had made an exaggerated report.

Stocks continued to fall for over a month, wiping out untold thousands who had bought on margins. San Francisco was in a turmoil. Few stopped to think that they had all gone mad and that the market had swollen to the bursting point. The actual losses were heavy, but are not to be measured by the quoted prices—much of the loss had been unreal "paper profits."

Bonanza stocks had not suffered greatly; Con. Virginia had fallen from $710 to $450; California from $790 to $250 (although quoted at $50 owing to the increase of 5 to 1 in shares). The disastrous declines had been in Ophir, from $315 to $65, and in all other Comstocks in which the general public had largely speculated.

The market touched bottom in the middle of February and began to climb. All thought the worst was over. The stock market was steady during the spring of 1875 and the Comstock had adjusted itself. The production from the Crown Point-Belcher and Con. Virginia bonanzas far exceeded the yield of all of the mines in earlier years. Con. Virginia gave added confidence by increasing its monthly dividends from $3 to $10 a share, or $1,080,000 a month, which was continued for thirty-four months.

Business was booming, the population increased faster than houses could be built, and the future of the camp was believed to be beyond estimate. Little did people dream of two other great calamities that were to be visited upon them that year—the failure of the Bank of California on August 26 and the great

[6]Quoted in Dan DeQuille's *Big Bonanza*, pp. 475, 476 (1876).

There was no savings bank in Virginia City, so Mackay organized one on August 26, 1875, with the following leading citizens as directors: C. C. Stevenson, J. C. Currie, L. T. Fox, C. Derby, C. E. DeLong, James Kelly, J. G. Fair, J. C. Hampton, G. W. Hopkins, Joseph B. Mallon, J. B. Hereford, George F. Hill, G. H. Winterburn, J. W. Mackay.

fire of October 26. The first of these was foreshadowed by the onslaughts of San Francisco newspapers upon Ralston and the Bank of California. The "Bulletin" especially was engaged in a crusade against the alleged unsafe management of the bank, which made little impression as Editor George Fitch was looked upon as a cranky, well-meaning man who was forever tilting at windmills.

THE CHRONICLE ATTACKS THE BANK CROWD

The San Francisco newspapers of the '70s were brilliantly edited, but, with the exception of the "Alta," were specializing in self-righteous attacks upon individuals, politicians, and corporations. Enterprising men lived in terror. Blackmail was rampant.[7]

The "Chronicle," following its usual policy, on May 20, 1875, turned its guns upon the Bank Crowd, blasting their management of the Ophir and denouncing their selfish and dishonest control of fifteen other Comstock mines whose trustees were their hirelings.[8] The newspaper was then boosting the bonanza mines and the managers and may have thought it was doing them a favor. As a matter of fact the exposé weakened confidence and the market.

The article, entitled "The Stock-Jobbing Juggernaut," takes for its point of attack the fact that the management of the Ophir, whose trustees were well known as the representatives of Sharon, Ralston, and Mills, had levied an assessment of $216,000 upon its shares at a time when the mine was producing 150 tons of rich ore every day. That, too, "at the moment the stock market is most disturbed, when margins are threatened, and fortunes hang upon a rumor."

The article says in conclusion:

> And now that we have described the temple of the stock-jobbing Juggernaut, its idols and its priests, given some hints of its interior workings, we can only say to our intelligent readers, do as you please. If you believe you can cope with these inside managers of mines successfully, do so. If you think you stand upon such equal terms in point of information as gives you an even

---

[7]*Alta California*, January 13, 1877.

[8]Neither Sharon, nor Ralston, nor Mills became members of the boards of trustees of the various mines which they controlled. Their representatives did as they were told.

chance with them to win at this hazardous business, go in; these persons will roll so long as you make the game. Dealing in mines under honest and honorable management is a hazardous business. Dealing in stocks under dishonest and dishonorable management will result in inevitable loss.

That attack was thought severe at the time, but it was mild compared with the later venomous assaults upon the Bonanza Crowd.

PARTIAL RECOVERY IN THE SPRING

Notwithstanding the collapse of the stock market in the early part of 1875 the estimates of the quantity of ore in sight in the Bonanza mines continued to be absurdly extravagant throughout the spring and summer.[9]

None of the mining men who made such fervid reports on the bonanza had given it a cold-blooded examination, based on a thorough sampling. All of them were stock gamblers and carried away by enthusiasm. The ore body was so irregular, so mixed with porphyry, and so slightly developed, that at best the result would be merely a good guess.[10]

The sampling of Caesar Luckhart,[11] a capable engineer, who reported for Rossiter W. Raymond in March 1875, resulted in the only reasonable report on the quantity and value of the ore then in sight in the bonanza mines, printed in "U. S. Mineral Resources" for 1874, pages 196–200:

> The 1500-foot level has, up to March 31, 1875, proved 700 feet length of ore, in some places 200 feet wide. The Virginia Consolidated is extracting an average of 580 tons of $150 mill ore per day from between the 1400- and 1500-foot levels. * * * But it is impossible to estimate the amount and value of ore standing in the Virginia and California ground. As has been remarked, the ore varies in quality. Extraordinarily rich streaks are met with. In some places for 50 feet in the crosscuts the ore assays $900, and again sections of 200 feet assay

[9]*Mining and Scientific Press*, July 3, 1875; *Virginia Evening Chronicle*, July 1875; *Big Bonanza*, p. 488; Powell's *Land of Silver*, pp. 73–101 (1875).

[10]Mackay said at the annual meeting of stockholders in January 1877: "It is impossible for us to tell two or three days ahead what the mine will develop."

[11]No other man was more familiar with all of the mines on the Lode. He had been Sharon's confidential engineer from 1867 to 1871; reporting to him daily the developments in the various mines.

$150. Crosscut No. 2 in the California, for 80 feet across the ore, assays $412 per ton, and, on the other hand, in some places the ore will not assay above $50; as, for instance, for 50 feet in Virginia Consolidated crosscut No. 2, near the west clay. Fabulous reports and estimates have been given, varying from $150,000,000 to $1,500,000,000 "in sight." It is reasonable to suppose that the ground from the Best & Belcher north line to the Ophir south line and in depth from the 1300- to the 1550-foot level, will yield, judging from present developments and present yield, ore for two years to come at the present rate of extraction, viz, 450 to 580 tons daily. The quality of the ore is a matter of pure speculation. At present the average is about $150 per ton, but $300 to $500 ore can be extracted at will, by selecting the stopes. Mr. Luckhart thinks that $160 may possibly be an average of the whole mass referred to, but this is merely a prudent guess. At this rate there would be over $60,000,000 "in sight." But such calculations are at present futile.

Raymond's report, issued in the spring of 1875, no doubt influenced the downward course of the market.

THE NEW YORK TRIBUNE CORRESPONDENT

The Director of the United States Mint, H. A. Linderman, thought the reported bonanza was of such national interest that he came to make a personal inspection in July 1875, accompanied by Prof. R. E. Rogers of the University of Pennsylvania, upon whom he relied for an expert report. The "New York Tribune," hearing of this proposed visit, sent on a staff correspondent whose excellent descriptive articles, beginning August 27, were as enthusiastic as the reports of the Director and his expert. Flood came up from San Francisco and he and Mackay and Fair showed the visitors every courtesy, which the correspondent reciprocated in praise of his hosts, their unassuming manners, their ability, and their devotion to the properties under their control.

The Comstock was at its best, and the correspondent was thrilled with everything he saw. No local newspaper man could have been more enthusiastic. "The mines in this city," he wrote, "are so marvelous in extent and operation that I should hesitate to give the facts in relation to them were they not certified by proofs so convincing that doubt is impossible."

His visit to the Con. Virginia was the highlight of his experience. At the assaying and melting department he saw bars of silver and gold stacked up like cordwood and "thrown about as if they were so many pigs of iron"; their weight ranging from 90 to 110 pounds and the value from $3,000 to $4,000 each. The trip through the steaming underground workings filled him with wonder—the hive of industry. The perfection of the arrangements, and the masses of ore everywhere were beyond anything he had imagined.[12]

The correspondent was unrestrained when he came to describe Virginia City:

> Here is a city of about 25,000 inhabitants, about 7,000 (6,200) feet above the level of the sea, with inhabitants in the garb of laborers, but with the habits of Parisians. Here are restaurants as fine as any in the world, though not so extensive as some, nor as elaborate in appointments; here are drinking saloons more gorgeous in appointment than any in San Francisco, Philadelphia, or New York; and here are shops and stores which are dazzling in splendor. The people here seem to run to jewelry. I have never seen finer shops than are here, and the number of diamonds displayed in the windows quite overwhelms one's senses. The Washoe Club is nearly as well furnished as any in New York, except in pictures, books, and bronzes, and the manner of living of the inhabitants generally is upon a high scale. * * * I have never been in a place where money is so plentiful nor where it is spent with so much extravagance and recklessness. * * *
>
> The houses are mostly of brick on the business streets, and the sidewalks swarm with people. It is as difficult to get along on C street in the evening as it is to go along Broadway in the neighborhood of Fulton in the middle

[12]Dan DeQuille wrote a more minerlike account of a trip through the Con. Virginia in the course of which he says: "Comstock silver ore is not as many may suppose—a bright and glittering mass. In color the ore runs from a bluish grey to a deep black. The sulphuret ore (silver glance) is quite black and has but a slight metallic luster, while what is called chloride ore is a kind of steel grey, but in places has a slight green tinge—the green showing the presence of chloride of silver. Throughout the mass of the ore in very many cases, however, the walls of the silver caverns glitter as though studded with diamonds, but it is not silver that glitters, it is the iron and copper pyrites that are everywhere mingled with the ore." Dan DeQuille's *Big Bonanza*, pp. 293–328, 336–353 (1876).

UPPER GOLD HILL IN 1875, LOOKING TOWARD THE DIVIDE (See bottom opposite page)

of the afternoon. Every young blood in the city, and every old one too, for that matter, has his fast horse or his pair * * *. I doubt if there is a city of 200,000 people in the United States which has as much wealth as Virginia City.[13]

The famous Washoe Club, of which the correspondent speaks, was formed by sixty prominent citizens on February 20, 1875. The Club's first luxurious quarters on "B" Street were burned in the fire that fall, after which it was permanently located in the Douglass Building on "C" Street. There the members, most of whom were mine superintendents, bankers, brokers, lawyers, and leading business men, gathered of nights in good fellowship— drinking, playing cards, and swapping yarns. Stories were current of poker games for high stakes. The register of the Club, upon which all of the noted men and women who visited the Comstock during the succeeding twenty years inscribed their names, is still on exhibition at Virginia City.

PROF. ROGERS AND THE DIRECTOR OF U. S. MINT OVERESTIMATE THE BONANZA

Professor R. E. Rogers' examination of the ore bodies in the Con. Virginia and California mines was made on July 16 and 17, 1875, and his report, dated November 13, was printed in the report of the "Director of the United States Mint" for the year ending June 30, 1875, on pages 81–83.

The Professor made his examination in two days, yet he speaks of it as a "careful and laborious investigation." He does not appear to have taken samples, but based his estimates on the mine maps which showed the quantity and the value of ore already extracted. It seems that he estimated only the ore between the 1400- and 1550-foot levels. The Professor's brief report to Director Linderman concludes as follows:

> On an inspection of the official surveys exhibiting the galleries and crosscuts, it would seem fair to conclude that with proper allowance, the ore body equals an amount which, taken at the actual assays, would give

[13]*New York Tribune*, August 27, 1875.

UPPER GOLD HILL—The Yellow Jacket hoisting works are at the lower right; Imperial-Empire hoisting works are at the top center. The long white dump of the Little Gold Hill group of mines shows at the upper left, beyond which lies the Bullion. The track of the Virginia & Truckee Railroad shows at the lower left. The hill in the background was known as Fort Homestead. A pleasure resort crowned it in the old days.

as the ultimate yield of the two mines $300,000,000, but to guard against a chance of overestimating, I take the assays at one-half that ascertained, which will place the production at not less than $150,000,000.

The Director in his report not only approves of Professor Rogers' estimate but makes the prediction that the Comstock mines will produce $50,000,000 a year for ten years.[14]   The "Tribune" correspondent reported that "When we fell in with Dr. Linderman (in the Con. Virginia) he was half crazy with excitement.   He thought it was time to resume specie payments, and was ready to begin at once."[15]

Squire P. Dewey, in his vicious pamphlet entitled "The Bonanza Mines," printed in 1878, charges that the Director of the Mint, Dan DeQuille and Deidesheimer were bribed with stock to make their extravagant reports.   It meant nothing to Dewey that Dan DeQuille was above suspicion and always broke, and that Deidesheimer went through bankruptcy a few months after the crash with liabilities $534,000.   The reader will be shocked at the suggestion that the Director and, necessarily, Professor Rogers, were bribed.

THE FIRM ORGANIZES THE NEVADA BANK

The overlordship of the Bank Crowd had been galling to the members of the Bonanza Firm, and as soon as the dividend on each of the 108,000 shares of the Con. Virginia was increased from $3 to $10 a month, they proceeded to organize their own bank, which they called The Nevada Bank of San Francisco.   "In former times the proprietors of the Con. Virginia patiently endured the domination of the Bank of California until they could act independently," wrote Myron Angel.[16]

The capital of the new bank was the same as that of The Bank of California, $5,000,000, gold.   A branch was immediately established in Virginia City across the street from the branch of The Bank of California.   A year later, when the deposits were $13,947,910, and cash on hand $5,495,712, the capital was increased to $10,000,000, gold, then said to be the largest capital of any bank in America.   Flood and O'Brien gave a check for $5,000,000

---

[14]Professor Rogers reports the principal metals in Con. Virginia ores are: gold (metallic), silver (metallic), polybasite, stephanite, blende, galena, and a small quantity of horn silver.   The Professor fails to include argentite, which was the principal silver sulphide.

[15]*New York Tribune*, August 27, 1875.

[16]Thompson and West *History of Nevada*, p. 595 (1881).

to pay for the new stock. There was no occasion for so large a capital and the increase appears to have been made for display.

The Nevada Bank was moderately prosperous until the summer of 1887 when, during the illness of Flood and the absence of Mackay in Europe, the then manager, George L. Brander, began to finance the attempt of Dresbach & Rosenfeld to "corner" the wheat of the Coast. It proved impossible, and Brander, in desperation, cleaned the bank of all of its resources. When Flood learned of the bank's condition he sent word from a sick bed to close the doors, but Tom Brown, cashier of The Bank of California, who dreaded a panic, advanced $1,000,000 in gold. Fair came forward with another $1,250,000, and the doors opened as usual on Monday morning. Fair with whom avarice was a passion, always kept a large sum of money on hand. He knew that his loan was safe, as Mackay and Flood were the sole owners. They had bought his shares two years earlier in order to get rid of him. That partially healed the breach, and Mackay on his return made him president of the bank to give assurance to the public, which was disquieted by false reports that Flood and Mackay had backed Dresbach & Rosenfeld. Fair never ceased to pose as a savior. On his election as president the "Chronicle"[17] commented: "Old Jimmy Fair has a good deal of the Indian blood in his composition, and he has waited long for a chance to get even on his old partners."

The Nevada Bank was a costly venture for Mackay and Flood. They paid Fair $2,000,000 for his interest in 1885, and Mackay said it cost them $11,000,000 more to restore the Bank and dispose of the 56 cargoes of wheat, then at sea, upon which Brander had loaned money at inflated values.[18]

Mackay's fortitude of spirit and business ability never shone more clearly than in that crisis. He was then in the midst of a rate war with Jay Gould over Atlantic cable tolls. Gould, who scented blood, pounced upon him in a malicious article in the "New York Tribune" of August 19, 1887, which said that he and Flood

---

[17]*San Francisco Chronicle*, September 14, 1887.

[18]When Fair was elected president the *San Francisco Chronicle* of September 14, 1887, printed a story of the "Wheat Deal" that is fairly correct. Seven years later, after Brander was indicted for making false reports as manager of an insurance company, the *San Francisco Examiner* printed the story under the captions: "George L. Brander's Confession That He Emptied the Nevada Bank. Inside Story of The Great Wheat Deal in Which Eleven Million Dollars Were Lost." (*San Francisco Examiner*, December 10, 1894.)

had financed the "Wheat Deal," that Mackay's fortune had been largely used up in his Atlantic Cable and Postal Telegraph ventures, and that the "Wheat Deal" would finish it. The plain purpose was to warn the business world not to assist Mackay. Fair is dragged into the article to say: "Poor John! most of his money is now at the bottom of the Atlantic."[19]

Mackay's instant reply was to reduce Atlantic cable rates from 25 cents to 12 cents a word; at the same time notifying customers who were under contract to pay 25 cents a word that their rates were reduced. That defiance and the restoration of the bank convinced Gould that it was useless to fight Mackay: "If he needs another million he will go into his silver mines and dig it out." Rates were then fixed on Mackay's terms.[20] The Nevada Bank was sold to I. W. Hellman in 1890 and lost its identity some years later in Hellman's Wells Fargo Bank.

THE BANK OF CALIFORNIA SUSPENDS ON AUGUST 26, 1875

Notwithstanding the incessant attacks upon Ralston and the Bank by the "San Francisco Bulletin," and to a lesser extent by the "Call,"—such things were expected of the San Francisco newspapers—no one suspected that both were on the verge of ruin, and that Ralston was making every possible shift to avoid it. Selling his stocks, turning over to his associates his one-third interest in the Virginia & Truckee Railroad, and large holdings of valuable real estate, overissuing bank stock and borrowing money upon it, withholding bullion belonging to the bonanza mines, and taking risks born of desperation.

However, there was something in the air during the two days preceding the suspension. A few cautious businessmen began quietly to withdraw considerable sums, but there was no sign of a run until the morning of the 26th which increased hour by hour until the doors were closed at 2:40 p. m.

---

[19]Fair believed in keeping his money where he could see it. He loaned money on mortgages and was known as the owner of more valuable real estate than any other man in San Francisco. Walsh said that Mrs. Fair loaned Mackay about $4,000,000 at this time in gilt-edge securities received on her divorce settlement.

[20]Dan DeQuille's *Big Bonanza* merely whetted Mackay's ambition. In 1883 he adventured into the Atlantic cable field, then controlled by a powerful monopoly. It was a gallant fight, and proved a profitable investment. The Postal Telegraph never became a successful competitor of the Western Union. Mackay's last and greatest commercial enterprise, the Pacific Cable, which called for an investment of $12,000,000, was in course of construction at the time of his death on July 20, 1902.

King tells a vivid story of the events leading up to the suspension and says the slaughter of prices occurred on the morning of the 26th, when Sharon gave the largest selling order ever known on the board.[21]

The average business man and people in general were stunned. The Bank of California had been the leading financial institution on the Coast for years. Many of the strongest men in the city were directors and stockholders, and it was regarded as impregnable.

The differences between the Bonanza Crowd and the Bank Crowd, as they were often called, were well known, and the proposed Nevada Bank was a threat to the supremacy of the Bank of California. The newspapers immediately printed the rumor that the Bonanza Crowd had not only withdrawn millions from the Bank of California, but had locked up several millions in order to embarrass it. Both Ralston and Flood promptly gave interviews to all of the leading newspapers in which they told of pleasant relations and denied the withdrawals. Nevertheless, some who pretend to write history, preferring sensation to fact, have repeated the charges. As a matter of fact, both Mackay and Flood admired Ralston and had no personal differences with him. Sharon they disliked, to put it mildly.

RALSTON'S DEATH

Ralston had been struggling for two years with the mounting obligations of his many enterprises.[22] He owed nearly $10,000,000, of which $4,000,000, unsecured, was due the bank. In normal times his assets would have been ample to cover all indebtedness, but not then. In desperation he had overissued 13,180 shares of

---

[21]King's *History of San Francisco Stock Exchange*, pp. 101–110 (1910). B. B. Rorke, Sharon's broker on the floor of the Exchange, "sold as long as there was a bid." Id., p. 106.

[22]The Panic of 1873 was in full swing in the East. Gold was at a premium and the West was being drained. From January to August 1875, the shipments of gold totaled $18,257,400; during the same period in 1874, they amounted to only $2,311,400. (*San Francisco Chronicle*, August 28, 1875.)

"For some time past the *San Francisco Bulletin*, and the *San Francisco Call* by their attacks upon the managers of the Bank of California, have been sowing the wind and now the entire business community is reaping the whirlwind." (Editorial in *San Francisco Chronicle* of August 27, 1875.)

The editor of the *San Francisco Call*, who had been pursuing Ralston, conceded on August 28: "It will be admitted that the Bank of California has done more to develop the resources of the Pacific Coast than all of the other banks in the city together."

bank stock and borrowed over $1,000,000 on it from various individuals and institutions. Reuben Lloyd said he thought "Ralston was worked out mentally. He had a great many things to do that he ought not to have been responsible for."[23]

At a meeting of the board of directors, his resignation as president was curtly demanded.[24] Nothing then remained but to care for his creditors as best he could. He made a new will in Sharon's favor, revoking one drawn two weeks earlier in favor of his wife; made an assignment of all of his property to Sharon as trustee for the benefit of creditors, then went alone to North Beach for a swim. Soon afterward his body was seen floating in the Bay, face downward. There was no sign of life after it was brought ashore. He was only forty-nine. His enemies said suicide; his friends, a stroke. In either case death must have been welcome. The coroner's jury brought in a verdict of accidental drowning, and the insurance companies paid his insurance, amounting to $65,000, to his widow.[25]

Sharon's ability was never more brilliantly displayed than in effecting the reorganization of the bank. He had a potent argument to bring other directors into line—their responsibility for declaring unearned dividends and for the condition of the bank. No doubt he repeated to them Reuben Lloyd's threat to hold them personally responsible.[26] Within five days after the doors closed

---

[23]Ms. in Bancroft Library (1886).

[24]A. J. Ralston said that Ralston's associates, "who had been greatly benefited by him, both directly and indirectly, failed to stand by him. They seemed absolutely incapable of a just and unselfish thought in the midst of disaster." (Ms. in Bancroft Library.)

[25]Cecil G. Tilton's *William Chapman Ralston* deals at length with those last days; also Julian Dana's recent biography of Ralston, entitled *The Man Who Built San Francisco* (1936).

"All of us living today who attended that funeral retain the kindest, warmest, most tender recollections of William C. Ralston." Joseph L. King's *History of the San Francisco Stock Exchange*, p. 110 (1910).

"Whatever the faults of Ralston, the people felt that he had intended to befriend the community in which he lived." *Bancroft's Works*, Vol. 7, p. 677 (1890).

Wells Drury aptly described him as "A man of vision, public spirit, and sanguine temperament." *An Editor on the Comstock Lode*, p. 118 (1936).

Ralston was portrayed generously and sympathetically by Francis G. Newlands, Sharon's brilliant son-in-law, in his argument in the Odd Fellows' Bank case, which involved some of the stock that Ralston had overissued. (*San Francisco Post*, November 13, 1885).

[26]Baldwin, who was in the East, had $1,800,000 on deposit, and Lloyd, his attorney, said he told Sharon that he and Mills and the other directors were personally liable to Baldwin, "and if something is not done I shall have to attach every piece of property you have in the world. * * * Then we all got together and arranged to reopen the bank." (Ms. in Bancroft Library, 1886.)

Sharon had secured subscriptions to the amount of $7,290,000 to restore its capital and provide for deficits. Sharon, Mills, Keene, and E. J. "Lucky" Baldwin subscribed $1,000,000 each, and sixty other loyal San Franciscans signed for the remainder.[27]

Much remained to be done, however, and the Bank of California did not reopen until October 2. The two leading stock exchanges, which had remained closed after the suspension, resumed at once.[28] Meantime the Nevada Bank Building was completed and the bank opened its doors for business on October 4.

Although stocks had been falling to new low levels every day in August, owing to a mysterious selling wave, the people of Virginia City were flocking to the Opera House, which was soon to go up in flames. Augustin Daly's Fifth Avenue Theatrical Troupe began a memorable engagement on August 16, 1875, and was followed by Lawrence Barrett in Shakespearean and other plays. On the night of the failure of the bank, Barrett played "Hamlet" to a large and enthusiastic audience.[29]

The closing of the branches of the bank in Virginia City and at Gold Hill was a blow to the people of the Comstock. Their money was locked up and their stocks largely sacrificed. But they were "mining camp people," accustomed to the reverses of fortune, and met the crisis hopefully. They still had their big bonanza.

Stocks opened at higher prices on October 5, when the stock exchanges reopened, and continued to climb until the day of the great Virginia City fire, three weeks later, when the exchanges were panic-stricken until they learned that the mines were not damaged below the surface.

VIRGINIA CITY DESTROYED BY FIRE OCTOBER 26, 1875

Confidence was restored as soon as it became known that the Bank of California was to resume, and stocks began to climb. Con. Virginia, the bellweather, which had fallen to $240 a share on the day of the bank suspension, rose to $320. Now everybody was going to be happy.

Then came the fire. A few minutes after six o'clock on the

---

[27]The list of subscribers is printed on page 397 of Tilton's *William Chapman Ralston*.

[28]A third exchange, the California, resumed ten days after the suspension and did an active business.

[29]*Daily Territorial Enterprise*, August 15, 1875; August 27, 1875.

Barrett made his first appearance in Virginia City in July 1868, and returned often in succeeding years.

morning of October 26, 1875, people were roused from their beds by the clangor of fire bells and the hoarse whistles of the mines, and within four hours the greater part of the city and the Con. Virginia and the Ophir hoisting works were in ashes. It was a bleak gusty morning, with a strong wind pouring over Mt. Davidson. There had been no rain for months, the wooden buildings were like dry grass, and the wind swept the town with a cyclone of fire.

The spectacle beggars description; the world was on fire. Almost a square mile of roaring flames leaping hundreds of feet in air. The down-rushing wind encountered the mounting flames and heat, causing great whirlwinds of fire and smoke that seemed to reach the heavens, bearing aloft showers of sparks and large burning fragments. The church bells clanged and mine whistles shieked until the fire silenced them. The continual explosions of giant powder gave an added touch of horror as men blew up rows of buildings in an effort to stop the spread of flames. The roof of the brick Catholic Church took fire, from which flaming shingles scattered abroad until a charge of giant powder blew the top off the building.

Goodwin tells the story that when the church took fire an old Irish woman rushed over to the Con. Virginia shaft where Mackay was working like a madman, and implored him to save it. "Damn the church!" he replied. "We can build another if we can keep the fire from going down these shafts."[30]   He made good that statement by contributing largely to its rebuilding.

The awesome scene reached its height when the fire caught the great Con. Virginia and Ophir hoisting works and millions of feet of piled lumber and thousands of cords of wood burst into flames. The heat was so intense that brick buildings went down like paper boxes "and railroad car wheels were melted in the open air alongside of the works."[31] Fortunately there had been ample time to get the men out of the mines.

Before noon the fire had burned itself out. On the north it reached the newly built First Ward School, on the south it was

---

[30]*As I Remember Them*, p. 161.

The author's mother collected a few belongings and led her four small boys up to the old Sides waste dump, from which we watched the fire.

[31]Ophir report for 1875, by Superintendent Captain Sam T. Curtis.

Wells Drury said of that fire: "It was my fate to be in the middle of flaming San Francisco in 1906, and the great Berkeley fire of 1923, but the intensity of that mining camp blaze lingers in my memory as the fiercest of all. Mt. Davidson seemed bursting in volcanic eruption." *An Editor on the Comstock Lode*, p. 120 (1936).

checked at the tall brick Odd Fellows Hall and the new Marye Building by blowing up adjoining buildings. The heart of the city, exceeding half a mile square, was utterly destroyed. The damage was estimated at $10,000,000, although half of that sum would be nearer the fact.

The Gould & Curry and the C. & C. hoisting works were saved. The only mining plants consumed were those of the Ophir and the Con. Virginia. The Bonanza Firm's fine new mill at the Con. Virginia was destroyed, together with the adjoining crushing plant of the California. The California pan mill, belonging to the Firm and then in course of construction, was below the line of fire.

Two thousand people were left homeless, many camping out in the sagebrush in the cold fall weather until they could be cared for. All thought was of them for a few days until they could be fed and clothed.[32] The people of the West responded generously. The women of Carson City promptly sent a carload of provisions and cooked food. Blankets, clothing, money, and supplies came from everywhere.

Every possible precaution was taken at the Con. Virginia by Mackay and Fair to prevent the fire from burning down the shaft. The cages were lowered a few feet below the surface, the safety clutches sprung, and dirt and ore piled on top. The mouths of the shafts were then covered with heavy timbers, and dirt and ore piled upon them.[33] Fortunately, the plan was successful. Although similar efforts were made at the Ophir the fire burned down the shaft for 400 feet.[34]

If the fire had reached the timber-filled stopes of the Con. Virginia the mine would have been practically ruined. Years would have elapsed before the stopes could be reentered; the workings would have caved, leaving ore and waste inextricably mixed. The fire that started in the stopes in 1881, after all of the rich ore had been mined, was only partially under control after four years.

As soon as San Francisco learned that the Con. Virginia and the Ophir works were on fire everybody began to sell and stocks crashed to new lows for the year. The Stock Exchange was an insane ward: "The discordant din was such as the walls of the board room had never before reechoed. Margins were swept out

---

[32]A quick-thinking uncle on Truckee Meadows, seeing the clouds of smoke, hurried up with a big hay wagon and rescued the author's family.

[33]*San Francisco Chronicle*, October 28, 1875.

[34]The Ophir report of December 15, 1875, by Sam T. Curtis, superintendent, tells of the efforts and the expense of rebuilding. Printed in *U. S. Mineral Resources* for 1875, pp. 147–150.

of sight in a moment. Dealers who thought themselves wealthy when they rose in the morning were paupers before 11 o'clock."[35] Con. Virginia broke $100 a share within an hour, a loss in market value of $10,800,000; California (with 540,000 shares) fell $10 a share; other Comstocks suffered still heavier losses proportionately.

The market recovered somewhat after Mackay reassured the public on the 28th in a statement to the reporters: "I have been through all the mines this morning and they are all right. There is no gas or fire in any mine connecting with the Gould & Curry." He regretted to state that the majority of the 700 men that had been employed by the Con. Virginia would be laid off until operations could be resumed in about sixty days. Meantime, he said, about 300 tons of ore would be hoisted daily through the Gould & Curry shaft.[36] The "San Francisco Chronicle" of the 31st, in repeating the interview, spoke of him as "the unassuming Mackay."

Two days after the fire the "Enterprise" concluded its editorial review of the situation: "Amid what looks as if it ought to be enough to cause universal despair, there seems to be a brave confidence and unflinching determination to overcome the present misfortune."[37]

FAIR'S MISLEADING REPORT FOR 1875

Superintendent Fair's report of Con. Virginia operations for the year 1875, published December 31, was inexcusable in its misrepresentations, and plainly designed to strengthen the market. It magnifies the ore bodies developed on the 1400, 1500, and 1550 levels, and suggests unexplored areas of great possibilities. Its worst feature is the statement that on the 1550-foot level, at a point 320 feet south of the north line of the mine, a double winze has been sunk to the depth of 147 feet, "all the way through ore of very high grade." This winze, in fact, was sunk only 160 feet south of the north line and in the southern edge of the ore body. The report, therefore, added 160 feet of nonexistent ore, and more

[35]*San Francisco Chronicle*, October 27, 1875.

[36]*Sacramento Record-Union*, October 29, 1875.

[37]*Daily Territorial Enterprise*, October 28, 1875.

The *Enterprise* building was destroyed, together with all of those precious early files. So far as known, there is no complete file in existence prior to April 4, 1866. The paper was printed in the office of the *Gold Hill News* until a new press could be obtained. The file beginning in 1866 is in the library of the University of Nevada.

than doubled the length of the ore body below the 1550 level at the south end, for, at the depth of 75 feet, the winze encountered the south edge of the ore body and was inclined northward in order to follow the ore, which was rapidly contracting in that direction. The report further says that work in the winze "has been temporarily discontinued on account of the increase of water and our limited means for hoisting."

Fair states also that "from this same level (the 1550) north of the northern line (in California ground) another double winze has been sunk to the depth of 128 feet through excellent ore the entire distance, and terminates in rich ore." The developments made by these winzes, he says, "prove the continuity at these lower depths of the same ore body which exists on the level above, with an appreciation in the quality of the ore."

Those reports were the subject of bitter criticisms in the years immediately following, when it became apparent that the ore body below the marvelous 1550 level became smaller and lower in grade. Dewey, who stopped at nothing in his attacks upon the Bonanza Firm through the columns of the "Chronicle" and in his pamphlet "The Bonanza Mines," found weapons for his hands in those reports. Not only Fair, but his associates suffered from his highly colored and misleading representations. Both winzes had passed through the ore on what became the 1600- and 1650-foot levels, and the decreased values must have been known.

Fair's report for 1876 recounts the rich ore in the winzes and the great expectations of the 1650-foot level, but omits the false statement that the Con. Virginia winze is 320 feet south of the north line.

The Comstock was astounded when Fair began to criticize Mackay publicly for "gutting" the Con. Virginia in the spring of 1876, during Fair's absence in the East. Fair's charges must have been wormwood to Mackay, but he gave no sign.

Fair's foxy methods were often illustrated in the old days by the story of his detection of miners smoking, and then laying the blame upon Mackay for their discharge. The rule against smoking in the bonanza mines was ironclad, owing to the dread of fire. One night as Fair was passing through the Con. Virginia workings he thought he smelled tobacco smoke. Putting on his most genial air he sat down near the men and whined, "Boys, this running around wears me out. I can't stand it any more. If I had the whiff of a pipe it would make me feel better." One was promptly offered and he took a draw or two, expressing his relief.

The next morning he happened to meet those men coming off shift, one of whom said they had been "given their time." "Ah! that's John again," sighed Fair; "I never get a good bunch of men that he doesn't lay them off." But the men knew better.[38]

[38]Judge Goodwin tells the story in *As I Remember Them*, p. 180 (1910). Daggett repeats it in *San Francisco Chronicle* of April 3, 1898.

CHAPTER XX

1876: High-Water Mark on the Comstock—A Troublous Year for The Bonanza Firm—The Market Revives and the Bears Threaten—Fair Claims Mackay Gutted the Con. Virginia— Keene Leads a Smashing Bear Attack on Bonanza Stocks— Mackay, Fair, and Flood in 1876—A Lively Con. Virginia Meeting—The Bullion Tax Fought by the Bonanza Firm.

The year 1876 was high-water mark in the history of the Comstock. Virginia City was rebuilding feverishly after the fire; more mines were in operation and more men employed than ever before; the population of the region reached nearly 23,000, including several thousand transients;[1] the Con. Virginia and the California were milling 1,000 tons of rich ore every day and each paying a monthly dividend of $1,080,000.

All of the large mines were sinking their incline shafts as rapidly as possible, and eight separate groups had joined in sinking third-line vertical shafts far to the eastward to intersect the Lode at depths varying from 2,500 to 4,500 feet. No one doubted that other great and rich ore bodies would be found on the deep levels. Mrs. Bowers' prophesies of bonanzas to be discovered fell upon willing ears. Most of the mills in the region were busy. The Bonanza Firm had twelve in operation; the largest being

---

[1]The State took a census in 1875, which gave the following facts concerning Storey County, whose population was almost wholly in the towns of Virginia City, Gold Hill, and Silver City:

Number of families	1,755
Number of dwellings	4,185
Number of white males	12,073
Number of white females	5,980
Number of colored males	88
Number of colored females	46
Number of Chinese males	1,254
Number of Chinese females	87
Number of males foreign birth	7,637
Number of females foreign birth	2,389
Total inhabitants	19,528

Note the preponderance of males, even at that late period.

Native born Americans made up two thirds of the population in the early '60s, while in 1875 two thirds were foreign-born. Gold Hill was largely a Cornish town, while the Irish predominated in Virginia City. "Americans furnished a majority of the foremen, bosses, engineers, firemen, carpenters, blacksmiths, and machinists."

the new Con. Virginia and California mills, which reduced 630 tons every 24 hours.

The control of the leading mines was no longer held by individuals or by a few men as in earlier days, but had passed into the hands of three groups: The Bonanza Crowd, the Bank Crowd, and J. P. Jones.

Joseph T. Goodman wrote of the period:

> In the flush days, the Comstock was one of the busiest and most picturesque sights ever witnessed. Immense hoisting works studded the mountain sides for miles with their huge dump piles and capacious ore bins; quartz mills were thundering and grinding in every available nook of the neighboring cañons; a continuous line of many-muled teams hauling ore, wood, and merchandise constantly crowded the streets and outlying roads, and in the town itself were such throngs of people as one would expect to encounter only in the heart of a great metropolis. The scene was made doubly animate by the prevalence of hope and high spirits. Everyone had money or felt rich with the ease with which money was to be acquired, and all met upon an equal footing, because one that was poor today might be rich tomorrow.

The railroad was gorged with trainloads of ore and lumber and general freight; C Street was so thronged with people and with vehicles as to be almost impassable. In the evenings the crowds filled the street as well as the sidewalks. Life in the saloons, gambling halls, and restaurants went on night and day, owing in part to the fact that shifts in all of the mines changed every eight hours. Despite the prevailing drinking and gambling the community was remarkably law-abiding for a rich and lively mining camp.[2] Nearly all of the violent encounters that writers love to tell about are decorated with "red paint."

Yet the great prosperity of the '70s came only in part from the

---

[2]"The number of saloons was out of proportion to the population, an inheritance from earlier days, but drunkenness was no commoner than elsewhere, and no one could complain that the town was not well behaved or that it was disorderly." (George T. Marye, Jr., in *From '49 to '83*, p. 130.) Marye was something of a Puritan, too. He went to Virginia City in November 1874, to assist his father in his stockbroking business, whose commissions ran from $1,000 to $3,000 a day in those times. George T. Marye, Sr., was an honored resident of Virginia City for many years, and became one of the wealthiest brokers on the Coast, following a rule never to buy or sell a share of stock on his own account.

production of ore; the expenditure of assessment money was a larger factor. Of the 135 Comstock mines quoted on the San Francisco stock exchanges in 1876 only three, the Con. Virginia, the California, and the Belcher, were paying dividends, and all of the others levying assessments. The Belcher paid its last dividend that year.

That period of remarkable productivity and of enormous expenditure was the golden age of the Comstock—a period glamorous and fascinating beyond description to those of us who participated in it, but soon to live only in memory.

The Market Revives and the Bears Threaten

The stock market recovered rapidly after the San Franciscans learned that the mines below were undamaged by the fire. Con. Virginia rose to $385 a share by December 2; California (whose shares had been increased 5 for 1 in May) was quoted at $61; Ophir at $58 had almost doubled in price; other Comstocks increased proportionately. Renewed buying, principally foreign, and active support of the market by Flood kept Con. Virginia well above $400 all winter.

Keene and some of the other sharp brokers and speculators thought Con. Virginia was too high and that Flood was quietly unloading. The two winzes that had been sunk below the 1550 level were still full of water. The management claimed that the water and heat prevented further work there until drainage and ventilation could be secured by a crosscut from the C. & C. shaft, but the brokers discounted that and planned a bear attack.

Fair Claims Mackay "Gutted" the Con. Virginia

Fair was in the East in the spring of 1876 and Mackay was in charge of the Con. Virginia. A savage bear attack on the stock, led by Keene, was in progress, which Flood in San Francisco was making every effort to withstand. Mackay (no doubt with Flood's approval and perhaps at his suggestion) sought to overcome the criticism that the mine was failing by making a record production of $3,634,218 during the month of March. However, instead of quieting the bears, that great production stimulated them to more savage assaults, for they were fully informed that the output had been made at the sacrifice of much rich ore.[3]

---

[3]During the big market of 1874–1876, the leading speculators and brokers in San Francisco had "representatives" (spies) on the Comstock, who kept

When Fair returned he did not hesitate to say that "Mackay gutted the mine" in order to make a great showing during his absence, and so stated to George D. Roberts and James R. Keene when they spent two days inspecting the bonanza mines, which they repeated to the press.[4]

Fair merely seized an opportunity for self-glorification during Mackay's absence.[5]   He knew that the rich ore on the 1500- and 1550-foot levels would be nearly worked out by the end of the year regardless of the increased production in March, and sought to place the burden on his associate.   D. O. Mills, who followed Comstock affairs very carefully during the '70s, said that the successful mining of the Con. Virginia ore body "was due to a considerable extent to Mr. Mackay's method of control and working out."[6]

The yield of the Con. Virginia for the year 1876 was 145,466 tons of an average value of $105.31, a total of $15,315,613, out of which $12,960,000 was paid in dividends.   But that was Con. Virginia's last great year.   Dividends were suspended during the first four months of 1877 on the claim that the shaft was in need of repair.   In 1877 production fell to $12,758,603, and thereafter the decrease was rapid.

KEENE LEADS A BEAR ATTACK ON BONANZA STOCKS

James R. Keene, supported by a group of powerful brokers, notably C. W. Bonynge,[7] Mark J. McDonald, and Jack McKenty opened a smashing bear attack on bonanza stocks in March 1876, claiming that the Con. Virginia was failing. Flood was outraged. This was his first experience of the kind, although far from his last.   He and Keene heatedly debated the merits in the newspapers, but the bears were well-organized and stopped at nothing

them informed by coded telegrams of all developments in the mines. All had miners in their pay, and were most ingenious in obtaining information.   King comments mildly on this practice in his *History of the San Francisco Stock Exchange*, p. 243 (1910).

[4]The *Stock Exchange* of August 1, 1876; *Evening Post*, reprinted in *Daily Stock Report* of August 2, 1876.

[5]Mackay was Commissioner of Mining at the Centennial Exposition that summer and later accompanied Mrs. Mackay and the two little boys to Paris, where she resided for ten years; then removed to London where her social triumph was complete.   Mackay, who disliked social functions, lived in New York.

[6]*New York Herald*, July 21, 1902.

[7]Bonynge and McKenty had been brokers in Virginia City in the '60s and had seen all of the other bonanzas play out.

in their criticisms. Flood's invitation to the public "to visit the mine and see for themselves" had no effect.

The "San Francisco Chronicle," which had showered praise upon the bonanza mines and their managers for nearly two years, printed a criticism of both in its issue of July 17, 1876, which further demoralized the market. The article was signed "Unknown," but appeared to be from a well-informed person, generally thought to be Squire P. Dewey, although he denied it. This was the beginning of a campaign of vilification that was to last for years. It was said that an owner of the newspaper brought the article to Flood's office to ask if it stated the facts. He was told that it did not, but not getting them to see what he wanted, he went off with it.[8]

The bears broke Con. Virginia from $440 a share on February 24 to a basis of $240 on July 13. The capital stock meantime had been increased from 108,000 to 540,000 shares and the market price on July 13 was $48 a share. California had fallen from $90 a share to $65. In July Flood gave up trying to sustain the market, at the same time stating that he and his associates had not sold a share of their stock.[9] The bears had taken a heavy toll of him for months by "shortening" bonanza stocks.

After Keene had wrecked the bonanza stocks, making a handsome profit, he made an examination of the mines and returned something of a bull, although quite reasonable in his statements to the "Evening Post."[10] He believed that prices would improve but added the warning: "But you never can tell about a market." While Keene praised the outlook for the California, he did not find conditions so favorable in Con. Virginia. Nevertheless, he thought the stock a good buy.[11]

Keene left for the East on December 20, 1876, to become a leading figure in the New York market for many years.[12] Prior to his departure he said, in an interview in the Stock Exchange, that the developments on the 1650 level of the Con. Virginia

---

[8]J. M. Walker, in *The Stock Exchange*, July 18, 1876.

[9]*Daily Stock Report*, July 18, 1876.

[10]Reprinted in the *Daily Stock Report* of August 2, 1876.

[11]Interview in *San Francisco Chronicle* of August 2, 1876.

"A distinguished party visited the bonanza mines today—W. H. Russell of the *London Times*, Mr. Prior of the *London Illustrated News*, Mr. Weed of the *New Orleans Pickayune*, and Howard Coit of San Francisco." *Virginia Chronicle*, June 10, 1876.

[12]Keene was employed by J. P. Morgan in 1906 to float the shares of the newly formed United States Steel Company, and did it very successfully.

should be favorable when the crosscut from the C. & C. shaft reached the ore body. He thought the bonanzas would be good for another year's dividends of $2 a month on the 540,000 shares in each mine, "and then the gamble in favor of further development would make the stocks a good speculation." He did not pretend to be a mining man and scoffed a little at the reports of the "so-called experts."

Keene's withdrawal from the bear market failed to stop his associates, who continued to hammer bonanza stocks relentlessly and drove them to new low levels. Fortune favored the bears; the mines and the market continued to decline beyond their hopes and they reaped a rich harvest.

The California came into production in 1876, when the yield was $12,505,320. Its greatest year was 1877, when $17,879,947 was produced, and $14,040,000 paid in dividends, at the rate of $1,080,000 a month. Early in that year, however the lower limits of the ore body were disclosed and the market price of the stock fell under the onslaught of the bears.

MACKAY, FAIR AND FLOOD IN 1876

Some of the characteristics of Mackay, Fair, and Flood are told by a correspondent at Virginia City in 1876:

> Mackay is an Irishman, about forty-five years old. He is well made, has a clear blue eye, and wears a light mustache. At almost any hour of the day you may see him along the mines, his trousers tucked in his boots and a rough felt hat pulled down over his forehead, paddling along through the snow from the Savage to the Hale & Norcross, thence to the Gould & Curry, and so onward to the Consolidated Virginia and California. He has none of the airs of a monarch. If you happen to know him he will say a few pleasant words and bid you good morning. His demeanor is invariably quiet and modest, and the most jealous eye can detect no bluster in it. He seems to be somewhat distrustful of his conversational powers, and thinks more than he talks. He is talked of some for the United States Senate, but I doubt his ambition. He is not an educated man and knows he is not. Then his business here requires and will continue to require all his time for the next ten years.
>
> Fair is a very different sort of a fellow. He is full of

bonhomie (the people hereabouts call it blarney), a good talker, good-looking, and particularly social. He is rarely seen afoot, but lays back in state in his buggy and mentally puts up jobs on the stock market. Very sly is "Slippery Jim," as his admirers call him.[13]

Two years later when the Firm's mines extended from the Sierra Nevada on the north to the Yellow Jacket at Gold Hill, "Mackay, when paying his daily visit to the mines, used to ride in a buggy drawn by an old gray horse, the entire outfit not being worth $50, while Fair, on similar business, had a man to drive him behind a fine team."[14]

Flood was the active business man of the Firm, and the Street soon recognized him as one of the most astute operators, of good judgment and discretion, and possessing one of the best business minds in San Francisco. He was stout of figure, of medium height, his sandy hair turning gray, and pleasant and earnest in conversation. His generosity was proverbial.[15]

While Flood was a quiet, unostentatious man he followed the fashion of other wealthy men of the period by building a baronial estate in the country and an imposing brownstone mansion in the city. Since "The Fire" of 1906 the latter has been occupied by the Pacific Union Club.

A LIVELY CON. VIRGINIA MEETING

Stockholders of Comstock mines were more interested in the market than the management, and annual meetings for the election of trustees were usually tame affairs, attended by few. But the Con. Virginia meeting of January 1877 was long and contentious, owing to the criticisms of the management by Squire P. Dewey who became the nemesis of the Bonanza Firm. If they could have known the plagues that he was to visit upon them during the next four years, they would have been glad to hand him the $52,000 that he claimed to have lost through misinformation given by Flood. Dewey was accompanied by Mr. James

---

[13]*San Francisco Bulletin*, March 6, 1876.

[14]Professor J. N. Flint in *San Francisco Call*, July 21, 1889. Professor Flint, a highly respected man, was principal of the Fourth Ward School for years, and County Superintendent of Schools as well.

[15]King's *History of San Francisco Stock Exchange*, p. 212 (1910). The Firm nvested its profits in San Francisco real estate, which was divided among the members in the latter part of 1876.

White, an English stockholder, whose applause of Dewey's points enlivened the meeting. Mackay and Flood were indignant and excited and gave the "San Francisco Chronicle" an opening for personal abuse. All of its earlier laudations were forgotten. Mackay was now a "bulldozer," and his attitude toward Mr. White "might have been expected from a scrub, who, raised from nothing, had attained to nothing except dirty purse pride."

The "Alta" of the 14th denounced the "Chronicle" for its misrepresentations of the meeting and defended Mackay:

> We believe that no other of the great millionaires of our times is so noted as Mr. Mackay for conduct so inconsistent with purse-pride, for simplicity of manners, cordiality of relations with old friends, no matter how poor, and strict attention to hard labor, for his work at the mine requires much physical exertion.

Mr. White remained in San Francisco for several months and learned the other side of the story. On Mackay's invitation, he made an inspection of the bonanza mines and mills, accompanied by his expert, and returned fully satisfied. "So explodes another 'Chronicle' canard," commented the editor of the "Stock Report."[16]

The attacks of the bears upon bonanza stocks and the management, together with the criticisms in the "Chronicle" and the continued decline in prices, brought about a revulsion of public sentiment that amazed the editors of the mining journals, who, together with the conservative old "Alta California," made a vigorous defense of the Firm.[17] The "Daily Stock Report," which had been somewhat critical in earlier times, came to the defense of the Firm repeatedly in the latter part of 1877, and said in its issue of December 14:

> Flood and O'Brien [for so the Firm was known in San Francisco] have made mistakes, doubtless—all men are fallible—but for general good management and efficiency as well as for fair dealing and integrity, we have yet to meet their equals in the whirlpool of the street.

THE BULLION TAX FOUGHT BY THE BONANZA FIRM

A tax on the bullion product of mines has always been

---

[16]*Daily Stock Report*, May 2, 1877.

[17]*Alta California*, January 13, 1877; *Daily Stock Exchange*, July 17, 1876; *Daily Stock Report*, December 14, 1877; *Mining and Scientific Press*, January 27, 1877.

denounced by miners as an outrage and a form of robbery. They defeated the first proposed State Constitution in January 1864, because it put mines on an equal footing with other property. The next Constitution contained a provision in accordance with William M. Stewart's views and was adopted overwhelmingly.

The first Legislature, which was dominated by mining influence, passed a law discriminating in favor of mines over other forms of property. The bullion only was taxable, and that at a preferred rate, after deducting $20 a ton for the expense of milling. The act was clearly discriminatory, but was not tested until early in 1867, when it was declared unconstitutional.

Sharon, who had come into control of nearly everything in Nevada, had a special session of the Legislature convened in 1867, which enacted a new law still further discriminating in favor of the bullion product of Storey County. Again, in 1871, the Legislature did his bidding, notwithstanding the protests of the ranchers and other property owners throughout the State who felt they were shouldering the burden of taxation. They paid taxes on the assessor's estimated value of their property, while by law the undersurface value of the mines was assessed at a figure equal to the year's proceeds from ore production less the costs, but allowing undue high-fixed deductions. This valuation was but a fraction of the value placed on the prosperous mines as judged by stock quotations.

In addition, there was the very unfair provision in the law that while the property owner in Storey County paid the regular tax rate of approximately $1.50 per $100 valuation, the mines paid a fixed rate on their indicated profit from ore production of only 25 cents per hundred.

By 1875, the indignation of the property owners had reached such a pitch that even the representatives from Storey County helped to enact a law placing the net profit valuation on the same rate of taxation as other property, after a deduction for the reasonable cost of milling. Sharon was a candidate for the United States Senate before that Legislature and kept his hands off. The rich ore in his mines was practically exhausted and he no longer cared to exempt that kind of property. "In fact, his interests now demanded a change of policy. A heavier tax would draw but little money from his mines and would increase the

Thompson & West *History of Nevada* devotes pp. 122–130 to the subject.

Storey County sinking fund out of which the V. & T. Railroad bonds were to be paid.[18]

While the tax, averaging about $250,000 a year, was but a fraction of the total product of the bonanza mines, the indignation of Mackay and his associates at this change in the law led them into making one false move after another. They took the question into court and went into politics, but, in the end, the companies were compelled to pay in full.

[18]Thompson & West *History of Nevada*, p. 126 (1881).

CHAPTER XXI

1877: The Bonanza Terminates on the 1650 Level—Warring Brokers and Speculators—Hard Times in 1877—The Decline Begins.

THE BONANZA TERMINATES ON THE 1650-FOOT LEVEL IN 1877

The absorbing question in the minds of the speculative public at the beginning of 1877 was whether the Con. Virginia bonanza would hold its own on the 1650 level, then about to be opened by a long crosscut from the C. & C. shaft.

The market had not recovered from the disastrous bear attack of the spring of 1876, and the lesser bears, after Keene's conversion and departure, had kept it flat throughout the latter part of the year. Bonanza stocks had fallen to a low point in December 1876, and still lower in the early part of the following January. Con. Virginia closed at $36 a share on January 6, and California at $41.50. In the middle of January, when the crosscut encountered rich ore, Con. Virginia rose to $52 and "the bears were stampeded."

Geo. D. Roberts, "Bob" Graves, and J. T. Goodman came up from San Francisco to examine the ore in the crosscut and made reports as glowing as those of early bonanza days.[1] Goodman thought the mine was looking and promising better than ever[2] and backed his judgment by buying 500 shares of Con. Virginia, which brought a heavy loss. Graves' stock in Con. Virginia and in California brought him $80,000 a month in dividends in 1876, but he continued to hold it in a declining market and saw his great fortune melt into thin air.[3]

It happened that the crosscut had intersected the richest ore on the level. Further development work showed the ore body to be 200 feet long and 90 feet wide, but split from end to end by a porphyry horse 30 feet thick.[4] A similar body in the California was separated from that in the Con. Virginia by a mass of porphyry.

The sudden contraction of the bonanza in size and value below

---

[1] *Daily Stock Exchange*, March 10, 1877; *San Francisco Chronicle*, March 11, 1877; *San Francisco Post*, March 12, 1877.

[2] *San Francisco Post*, March 12, 1877.

[3] Marye's *From '49 to '83 in California and Nevada*, p. 100.

[4] This "horse" was mined during the low-grade period and yielded $16 a ton.

the 1550 level is explained by Plate 5 of the Atlas accompanying Becker's report, which shows a horse of diorite projecting upward into the ore body to the 1600-foot level and splitting it into two prongs of ore of diminishing size.

It is an open question whether the bonanza was formed in a shattered lens of diorite, which became largely replaced by ore.[5] The large lenses shown below the bonanza in Becker's Plate 5 are of diorite and Plate 8 shows that all of the deep workings in the Con. Virginia and the mines northward below the 1800-foot level were in the shattered diorite along the northeast rift. Strange to say, none of those rich little disconnected ore bodies in the rift were encountered in those early workings, nor until they were pumped out after 1900. The Lode itself was not prospected below the 2500-foot level. It had closed to a mere contact 1500 feet above that, and the mineralizing agencies escaped into the rents in the hanging wall.

WARRING BROKERS AND SPECULATORS

There were three active stock exchanges in San Francisco during the flush '70s, each with a large membership. Able and daring men gravitated to them by instinct. They were the liveliest places on the Coast and the focus of public attention.

The big speculators, like Sharon, Flood, Hayward, Jones, and Skae had their favorite brokers, who in turn usually employed other brokers to buy and sell on the exchanges. These groups were almost constantly at war with one another. Other large dealers, "Lucky" Baldwin for example, played a lone hand against everybody.

King says that prominent brokers in those days made more money than a bank.[6] "In flush times the leaders of stock operations were known by their purple and fine linen, their splendid equipages, and their lavish expenditures, generally in San Francisco, but sometimes in a trail of coruscating glory across the continent."[7] When "Jot" Travis, a well-known westerner, was being shown about the waterfront of San Francisco in the late

---

[5]Prof. John A. Church, whose investigation of the Comstock was made in 1877, wrote that the "horse" that separated the two ore bodies on the 1400-foot level was of diorite. (The *Comstock Lode*, p. 110.) Becker's examination was made after all of the bonanza ore had been mined and he does not speak of diorite in the stopes. "Below the bonanza," he states (page 270), "large masses of diorite are embedded in undeterminable vein matter and diabase."

[6]King's *History of San Francisco Stock Exchange*, p. 249 (1910).

[7]Shinn's *Story of the Mine*, p. 152 (1896).

'70s, and one fine yacht after another was pointed out as belonging to a wealthy stockbroker, he finally inquired, with an air of childlike innocence, "W-where are the c-customers' yachts?"

The underlying purpose of both brokers and speculators was to keep the public interested, as that was the fountainhead of the market, but in their day-by-day operations they watched one another like hawks, trying to guess the play and assisting or blocking every move as seemed to their advantage.

Flood succeeded Sharon as the dominant factor on the Board with the rise of the Con. Virginia and the control of adjoining mines, and was followed and flattered like a monarch. Life must have been sweet to him from the time the Firm took control of the Hale & Norcross in 1869 until the bears began to slaughter Con. Virginia in the spring of 1876. Then troubles began to multiply. He was beset on every side and scarcely knew a day free from care thereafter. He took the brunt of the stock market for the Firm and met the often scandalous and unfair criticisms face to face. Both he and Mackay winced under them but Fair and O'Brien were unruffled.

The termination of the bonanza on the 1650-foot level, coupled with the continued criticisms, apparently caused the members of the Firm to lose all interest in bonanza stocks, except to unload more of their holdings, and the stocks drifted downward as the ore became depleted and the dividends grew smaller and smaller. They retained control of the mines, however, with the aid of their friends, and continued the search for another bonanza on level after level down to the 2,900, which was connected by a large drift with the Ophir, Mexican, Union, and Sierra Nevada.

The members of the Firm were also hopeful that other great ore bodies were to be found in the shattered northeast hanging wall country in the mines adjoining on the north, and, in 1877, while continuing the development of Con. Virginia and the California, they bought the control of the Ophir and the Mexican from Sharon. A year later they took over the Union and the Sierra Nevada. Extensive and fruitless development was carried on in those mines for seven years until the joint Ophir-Mexican winze reached the depth of 3,360 feet, in barren diorite, whereupon pumping ceased on January 1, 1885.

Meantime, they acquired control of the Yellow Jacket in 1876 and sank the New Yellow Jacket shaft to the depth of 3,060 feet, at the cost of $2,000,000, without finding a ton of ore. Their last disastrous venture, which brought a loss of $5,000,000, was the purchase of the control of the Union and the Sierra Nevada in 1878 on Fair's recommendation, during the childish "Sierra Nevada Boom."

One marvels at the ambition of Mackay and Flood to take over additional mines and enterprises and load themselves with new burdens, for they took infinite pains and delegated little authority. Neither appeared to know the limits of flesh and blood. The desire was not so much to pile up more millions as to exercise the powers that had come to them after years of obscurity. They had to take their pay in that coin, for there was little happiness after the first flush of the development of the bonanza. The pride and satisfaction of discovering and managing the greatest mine in the world at that time were canceled by the floods of criticism and the litigation that followed.

Flood and his partners guessed the market wrong at every important period. They did not know the ways of a stock market as well as Keene and other brokers who so successfully raided bonanza stocks. Nor were they as well able to judge of the merits of their bonanza—they were too close to it. Flood bought and sold the stocks of the mines controlled by the Firm[8] in a large way with the advantage of "inside information," such as it was. The advantage was slight, for the mines were open to the public except during brief periods. Besides, concealment was practically impossible, so complete was the spy system of the other large operators. Again, all moves had to be made on the floor of the Stock Exchange where every play was outwitted as far as possible. It is a maxim of Wall Street that inside information will break anybody, because, as Keene said, "You never can tell about a market. * * * Stocks have gone up when insiders were bearish and gone down when every director was an enthusiastic bull. A man may know his own business and not the stock market." In short, Flood's undoubted talent for stock-market operations was not profitable in the end, and brought only a world of cares. It was his misfortune to be supporting a failing bonanza and a declining stock market. Both Mackay and Flood, in later years, often expressed regret that they had not confined themselves to mining.[9]

Fair resigned as superintendent of all of the Firm's mines in July 1878, owing to ill health. Enormous labors in the mines and

---

[8]Flood confined his activities to their own stocks; he did not interfere with other men's game.

[9]James E. Walsh; R. V. Dey.

equally enormous appetites made him an old man before he was fifty. He ate ravenously, drank heavily on the sly, and his passion for women was so notorious that his good Catholic wife divorced him on the ground of "promiscuous adultery," two years after his election as Senator.[10]

Fair had little to do with the mines after advising the costly purchase of Sherwood's stock in the Union during the "Sierra Nevada Deal" in September 1878. Shrewd, and without sentiment, he thought the continued assessments levied to carry on deep mining were a waste of money, and, in 1881, as soon as the settlement of the Dewey suits permitted, he withdrew from the Firm, leaving only Mackay and Flood. O'Brien had died in 1878.

Fair's health continued to decline in 1879, and in 1880 he made a trip around the world with Dick Dey. On his return, much improved, he was waylaid by the leaders of the Nevada Democrats, who never had a "sack" at their disposal in a senatorial campaign and never elected a candidate. Fair had taken so little part in politics that he "did not know what party he belonged to." Nevertheless, the proffered "honor" appealed to him. Sharon, whose term was expiring, was the Republican candidate and "willing to go the limit." "In the entire history of Nevada politics there had never been such a saturnalia of corruption," wrote Sam Davis, who was then editor of the "Carson Appeal."[11] The "Gold Hill News" commented, "Now we shall see the open sacks of Sharon and Fair." Sharon was snowed under, and Fair had the good sense to become a silent member of the Senate.

The title of "Senator" was Fair's crowning vanity. He boasted that it cost a fortune. His health improved and his excesses continued until his death from a complication of diseases on December 28, 1894. He was only 63. To the last his self-approval was perfect. The traits that other men criticised he regarded as virtues. Fair's characteristics clearly appear in the reports of George H. Morrison to Hubert Howe Bancroft, leading up to the biography of Fair which appeared in Bancroft's "Chronicles of the Builders," Vol. 4, pp. 209–236.

Those reports, which include memoranda dictated by Fair in 1888, are among the treasures of the Bancroft Library. It would seem that the editor who dished up that laudatory biography carefully avoided reading the reports.

---

[10]Mrs. Fair and Mackay had been friends ever since he met his future wife at her home in 1867. When Mrs. Fair died in 1891, Mackay and Dey were appointed executors of her will.

[11]Davis' *History of Nevada*, Vol. 1, pp. 422, 423.

Morrison, who had known Fair on the Comstock, deals shrewdly, although not unkindly, with him, and says in conclusion: "He is no doubt a man of great ability. A very cool man, no conscience to trouble him, no sentiment. From the few interviews I have had with him I look upon him as the most remarkable man on the Coast." Harry B. Hambly, who was in Bancroft's employ in the '80s and '90s, told this writer that Fair paid $15,000 for the biography, and that he collected the last $5,000. Vanity was the only thing that loosed Fair's purse strings.[12]

HARD TIMES IN 1877—THE DECLINE BEGINS

The complaint of hard times on the Comstock during the height of the big bonanza sounds strange, yet the newspapers constantly dwell upon the lack of employment for many that had flocked in. As early as the summer of 1875 the "Enterprise" warns outsiders that there is already a large surplus of miners: "Mining bosses and superintendents are fairly harassed by these men, eager to get a chance to work, and the same is true of all kinds of labor."

There was little complaint during 1876 when the town was rebuilding and all of the mines in the region were in operation, but conditions became serious early in 1877. The "Virginia Chronicle" says that the destitution on the Comstock

arises from the very simple fact that there is a surplus of unemployed people waiting for developments in the mines which shall cause a demand for labor; many of them men with families. The Relief Committee has been disbursing money during the past few weeks. * * * The hard times are beginning to tell on everybody, high

---

[12]Fortune favored Fair while he lived, but the scandal and disgraceful litigation that followed kept the courts in a turmoil for almost seven years. Valuable papers were stolen from the public records; a respected school teacher, claiming to be Fair's "common-law wife," produced a penciled will and deeds to valuable property, all of which the courts declared fraudulent; and, finally, after decisions both ways, the carefully prepared trust provisions of his will were set aside by the Supreme Court in a four to three decision (132 *California Reports,* pp. 523–582). The contest was maintained by Fair's children who wanted to enjoy the estate instead of waiting on the provisions of the trust.

The lawyers never had such rich pickings, over twenty-five of whom took a hand at one time and another. Shuck tells of that litigation in "The Celebrated Fair Will Case." *History of Bench and Bar,* pp. 335–345 (1901).

The Justice who changed his vote and brought about the final decision resigned years later when accused of receiving a large bribe. *San Francisco Call,* November 22–25, 1918; June 6–15, 1919.

and low. Even the billiard halls and the gambling establishments complain. California Street is plucking the Comstock goose of even its pin feathers. * * * Nevertheless, people say, "It is always darkest just before dawn."[13]

Everything had been overdone in the flush year of 1876, and when winter came two or three thousand people remained without work and without means. Another depressing factor was the stock market, which had fallen to the lowest point in years when development work disclosed, early in 1877, that the big bonanza had practically terminated on the 1650-foot level. One hundred and thirty mines were levying assessments, and only the Con. Virginia and the California paying dividends. Some of the lesser mines began to curtail operations, while in the others, which were rapidly sinking their shafts, fewer miners could be used. Nevertheless, there were 3,156 men employed in and about the mines on the main Lode at the end of July 1877,[14] two thirds of whom were working in or about the ten mines controlled by the Bonanza Firm.

In November 1877 came the heartening news that the Ophir had struck ore on the 1900-foot level in an easterly and westerly fissure in the hanging wall, something like the one which led to the Con. Virginia bonanza. The hope that another bonanza was about to be developed threw the public into a fever of excitement for a time. Earlier in the year the Bonanza Firm had bought control of the mine from Sharon, who had found only some small ore bodies, and the new discovery looked for a time as if luck was again with the Irish. There was no concealment, and mine superintendents and newspaper men hastened to make an inspection, all of whom made favorable, although not extravagant, reports. Dan DeQuille, who accompanied the experts, reported that the vein was from 10 to 12 feet wide and that the ore ran about $40 to the ton, and was more or less mixed with porphyry and waste.[15] Fair immediately assumed the superintendency. If there was to be any glory he wanted it.

The new find, called the Hardy vein after Superintendent William Hardy who made the discovery, proved disappointing on

---

[13]*Virginia Evening Chronicle*, February 13, 1877; California Street in San Francisco; *Mining and Scientific Press*, May 12, 1877.

[14]The census of 1880 shows that the number of men employed underground had been reduced to 1,966, of whom only 394 were Americans.

[15]*Daily Territorial Enterprise*, November 12, 1877.

VIRGINIA CITY IN 1878

(See bottom opposite page)

the 1900-foot level, and the hope of another bonanza gradually faded. However, a rich, narrow ore body was developed between the 1900- and 2200-foot levels, that yielded 20,000 tons of $66 ore during the next two years, from which two dividends of $108,000 each were paid, the first from the Ophir in fifteen years. Again the mine went on the assessment list.[16]

During 1878 newspapers continued to speak of the necessity of helping the destitute, but the strain began to lessen in the latter part of the year when men and families began to leave in considerable numbers, especially after the "Sierra Nevada Deal" blasted so many hopes. They scattered far and wide, but it was the old discredited camp of Bodie that drew the larger number, after the little Bodie mine ran into rich ore in a drift on the 300-foot level in August 1878, and produced $600,000 in gold and silver in six weeks. The ore was so heavy with gold and silver that the metal caked under the stamps. The camp was twelve miles south of Aurora and shared in its boom in the early '60s, and likewise in its fall, as no profitable ore had been found. The inhabitants dwindled to a mere handful until 1876 when John F. Boyd and Seth and Dan Cook of San Francisco bought some claims for $60,000 in which excellent ore had been disclosed by a cave in an old incline shaft sunk in the early days.[17] The prospectors who found the ore were afraid to dig it out for fear of spoiling their mine, so they waited for a buyer. The purchasers organized the Standard Consolidated Gold Mining Company which built a small mill and straightway began to prosper. A small town grew up, but there was no boom until the sensational

---

[16]Some low-grade ore was found near the old surface workings of the Ophir in 1882, which has provided some of the chroniclers with a sensational story of surreptitious profits. A Cornishman named Rowe who had a house on the old dump found the ore while digging a cellar. He knew it belonged to the company so he mined it covertly and had it hauled away at night and reduced in a small mill down the Six-Mile Cañon. It was not long before Superintendent Patton heard of it and took possession. His report shows that he milled 4,268 tons from "the Rowe find" yielding $89,638.19 or $21 a ton. He then did considerable development work in the vicinity of the old Ophir workings and milled 8,578 tons, yielding $12 a ton, which involved a loss. Rowe's profits must have been small.

[17]John F. Parr, who was there at the time.

VIRGINIA CITY IN 1878—The northern part of Virginia City in 1878, looking toward the Flowery Range. This is the rebuilt portion of the town which was burned in the fire of October 26, 1875. The large white building above the International Hotel is Piper's new opera house. The Con. Virginia works are below the hotel, and the Ophir hoisting works stand north of the Con. Virginia. The cemetery and the Sierra Nevada works beyond.

CENTRAL PART OF VIRGINIA CITY, 1878   (See bottom opposite page)

discovery in the adjoining Bodie mine startled the West into another excitement.

Perhaps a thousand people left the Comstock for the new camp in the latter part of 1878, followed by more than double that number the next year. Other thousands flocked in from everywhere, among them many desperate characters, and Bodie suddenly became one of the largest camps on the Coast—and as reckless and bloody as Aurora in 1863. "The bad man from Bodie" became a byword. Fifty new mines started up, financed by assessments, and the camp was a whirlwind of excitement until 1881, when it became apparent that there were but two paying mines in the district, the Standard and the Bodie, both located on the top of Bodie bluff. A swift decline followed, although when the camp settled down it enjoyed a fair measure of prosperity for nearly fifteen years. Meantime the Standard acquired and worked out the adjacent mines, including the Bodie. In all it produced about $14,500,000, and paid about $5,000,000 in dividends.

The Standard mined ore from many parallel veins that dipped west and terminated against a north and south fault. No ore was found below 800 feet. Although the Bodie workings closely adjoined, the veins and the ore were of a different character, due to a change in formation. Its production did not exceed $4,000,000.

CENTRAL PART OF VIRGINIA CITY, 1878—The central part of Virginia City in 1878 from the water flume, looking eastward and down Six Mile Cañon. The rebuilt Catholic Church marks the southern limit of the great fire of October 26, 1875. The tall brick Odd Fellows Building which stands southwest of the Church was not burned. The Gould & Curry hoist and large white dump are at the right. The large building in the middle distance is the Sisters' Hospital. The C. & C. hoisting works are at the lower left with the top of the Pan Mill showing below.

CHAPTER XXII

1878 to 1881: Dewey Sues The Bonanza Firm—The Sierra Nevada Deal—General Grant and President Hayes Visit The Comstock—End of the Bonanza Period.

The sudden switch of the "San Francisco Chronicle" from fulsome praise of the Con. Virginia bonanza and its managers to a course of scandalous vilification was told in earlier chapters, which told also of the part played by Squire P. Dewey in those attacks. Both had lost money in Con. Virginia and both purposed that the Bonanza Firm should make restitution.

"Blackmail!" was the reply. The members of the Firm had never paid blood money and swore they never would. The discovery of a rich mine is always followed by litigation, chiefly blackmailing suits, and the Con. Virginia and the California had already defeated six actions of that kind. Flood and Mackay believed they had dealt fairly by their stockholders and scorned the thought of buying off the "Chronicle" and Dewey, although after five years of venomous newspaper abuse and two years of litigation with Dewey they wearied of the strain and compromised. The echoes of that barrage of unscrupulous and largely unwarranted criticism may be heard to this day—so deathless is slander.

No allowance was made for the fact that the bonanza was failing and that no new ore bodies had been discovered on the Lode, nor for the fact that the fortunes of all of the members of the Firm had become greatly reduced by the decline. No credit was given for their continued efforts to find new ore bodies.

An example of the "Chronicle's" abuse was an article five columns in length, which appeared in the issue of March 31, 1878, under scare headlines: "The Bonanza Kings. Their Splendor Throned on Human Misery. Rolling in Wealth Wrung from Ruined Thousands. California and Nevada Impoverished to Enrich Four Men. Plain History of Swindling Perpetrated on a Gigantic Scale. Colossal Money Power that Menaces Pacific Coast Prosperity."

This brought a heated editorial from the "Enterprise" denouncing the "Chronicle" and its owners, the de Youngs, with having reared "a pyramid of infamy on a foundation of blackmail."

Those owners have conveniently forgotten the fact
that the "Chronicle," after a careful survey of the

bonanza by its chief (Charles de Young), advised everybody that it contained limitless wealth—at a time too when the stock was selling at $600 a share.

The times were ripe for such attacks. Stocks had fallen year by year, leaving uncounted thousands ruined and resentful. The "Panic of 1873" had reached the Coast—there was no market for farm produce, manufactures were at a standstill, and times were desperately hard. Everybody was looking for a scapegoat. All of the ills that afflicted the people of California during those years were charged to Mackay and his associates. Some small newspapers joined the "Chronicle's" hue and cry.

Dewey did not bring his suits against the members of the Firm until the "Chronicle" had stirred up widespread public sentiment. The "Alta" and the "Bulletin" continued to denounce the "Chronicle" as a blackmailing sheet, but the latter was telling the people what they wanted to hear.[1]

While Dewey was working hand in glove with the "Chronicle" he was seeking in devious ways (as he admits in his pamphlet) to get back the $52,000 that he claimed to have lost because of misinformation given him by Flood.

Dewey, a self-important man who had grown rich dealing in San Francisco real estate, thought he was entitled to special information, and asked Flood if the dividend would be paid following the fire in which the Con. Virginia works were destroyed. Flood answered that the board would decide that at its next meeting, but referred him to the secretary for information as to the cash on hand. The secretary told Dewey the amount of cash, but failed to mention the bullion on hand. Dewey assumed that the dividend would be passed and sold a large block of stock, thinking the price would fall. The dividend was paid and the stock rose. Whereupon Dewey claimed that Flood was responsible for the loss he took in selling his stock.

When Dewey's efforts to get back his money had failed, he transferred 100 shares of Con. Virginia to John H. Burke, said to be "a recent arrival from the East, who has been a 'Chronicle' reporter for some months,"[2] who, on May 18, 1878, brought suit in his own name on substantially the same complaint in Dewey's

---

[1] *Con. Virginia Annual Report*, January 9, 1879. Dewey fills pages of his pamphlet with the *Chronicle's* articles—the name of the newspaper being omitted in many cases.

[2] *San Francisco Bulletin*, May 25, 1878.

name that had not been filed, but which he had handed to Heyden-feldt two months earlier with the suggestion that the matter be compromised by the payment to him of $52,000.

Burke's suits were brought as a stockholder of the Con. Virginia in behalf of all stockholders, to compel the trustees to disgorge $4,000,000 made in unlawful lumber profits, $26,015,000 in unlawful profits from milling and $10,429,068 alleged profits from the sale of the Kinney claim to the Con. Virginia.[3] It is noteworthy that while all other stockholders were invited to join in the suits, not one came forward.

The suit made a big stir in the newspapers. The editor of the "San Francisco Bulletin" deplores the fact that "blackmailing has long been a disgrace to our society," and suggests that:

> The firm of Flood & O'Brien, or any other prominent firm, will confer a lasting benefit on the community by fighting it to the bitter end. * * * If they set a good example now there are grounds for the hope that this miserable system can be torn up by the roots.[4]

A flood of ill-advised abuse was at once turned loose upon Dewey and his attorneys and upon the owners of the "Chronicle," which, of course, availed nothing.

Meanwhile the "Chronicle" continued its barrage of curses, and Dewey issued a 78-page pamphlet, entitled "The Bonanza Mines of Nevada—Gross Frauds in the Management Exposed," in defense of his own course and as an indictment against the Firm. Copies were mailed to all of the stockholders of the Con. Virginia and the California. The "Chronicle" of December 8, 1878, printed it in full, with a bitter editorial which concluded: "The whole history of this bonanza deal * * * is a history of duplicity, fraud, and cunning venality without precedent or excuse of any kind."

The pamphlet was a clever piece of evasion and misrepresentation, and so convincing to the uninformed that it has been quoted as authoritative by some writers, especially by the authors of two recent books of no value.

The first of the complaints to come to trial was the charge that

---

[3]The profits made by the Firm from the sale of lumber and firewood, as shown by the Mackay & Fair Letters, was $645,030. The milling profits amounted to $9,070,726.47. Another statement of milling profits, signed "W. H. L.," who says he compiled it from the Mackay & Fair books, places the total at $8,813,641.72.

[4]*San Francisco Bulletin,* May 25, 1878.

the Firm had defrauded the stockholders of the Con. Virginia of $10,429,068 by the sale of the Kinney claim to the company. The case was tried in San Francisco before Judge Jeremiah Sullivan, and the testimony was printed in full in the "San Francisco Bulletin," beginning December 15, 1880. The south boundary of the Kinney claim, which had but 50 feet of the Lode, was only 20 feet from the Con. Virginia shaft, and before work was begun by the Firm in the spring of 1872, it was thought necessary to acquire it.

The decision was printed in the "Bulletin" of March 30, 1881. The court found that the transfers were spread on the minutes of the board of trustees in April 1872, that there was no concealment, and that all who conveyed interests in the Kinney received treasury stock of the Con. Virginia in the same proportions, but held that it did not appear that Heydenfeldt was acting for Flood when he bought $12\frac{1}{2}$ feet from Kinney.

The court held that while the transaction was not "actually or willfully fraudulent," nevertheless under the rule of law then prevailing in California,[5] a director was forbidden to deal with the corporation for his own benefit regardless of the fairness of the transaction, and that he was liable to the stockholders for any profits resulting therefrom.

Judgment was thereupon entered against Flood and associates for the profits resulting from the $120\frac{1}{2}$ shares taken in Heydenfelt's name, amounting to the sum of $930,000, less the sum of $3,573 which Flood paid for the Kinney interest.

The judgment was for the benefit of all of the stockholders of the Con. Virginia, and the decree allowed them sixty days in which to appear and claim their share, less the expenses of the trial. The only one to come forward to share in the plunder was John L. Noyes.

It is probable that the case would have been reversed on appeal, but all of the parties were tired of the strife and the publicity, and after the sixty days had expired a compromise was made and all of the litigation terminated. The terms of the settlement were not made public, but we may believe that a sum far less than $930,000 was paid. John H. Burke, who appeared as plaintiff in the actions, wrote of the settlement some years later:

> With the consent of Mr. S. W. Holladay and John Trehane, my attorneys, and Squire P. Dewey and John

---

[5]The rule was changed by law in California in 1933.

L. Noyes, the only stockholders who had shown any dis-
position to join in the action, I settled the cases with
Hall McAllister, attorney for the defendants, much
against my desire. The full terms of the settlement
are private.[6]

THE SIERRA NEVADA DEAL

The "Sierra Nevada Deal" was the last big stock excitement,
and the one that broke the hearts of the people of the Comstock,
although their losses were trifling compared with those of the
gambling public in San Francisco. Always before they had taken
their losses cheerfully, for losses are inevitable to persistent gamblers, but this left them hopeless. It had been their last chance.
Not only that, but they felt they had been duped, which was true.
But the strange thing about it was that the big fellows, including
Jimmy Fair, had been fooled also, or, rather, had fooled themselves.

The booms of 1863, 1872, and 1874 grew out of great discoveries, but this almost incomprehensible madness of 1878 had for
its basis a small body of ore, which was so slightly developed that
no man could say there was $500,000 "in sight" at the very height
of the boom, when the market value of the Sierra Nevada reached
$28,000,000, and that of the adjoining Union had risen to
$18,200,000.[7] The former, which had 100,000 shares, had sold
for $2.80 a share in June of 1878, and reached $280 on September 27. Union had risen during the same period from $3 to $182.
The stocks of other Comstock mines had more than doubled in
market price; some of them, through manipulation, to five times
their selling prices in the spring.

That excitement can be explained only upon the theory of mass
emotion arising out of wishful thinking. They were all stock
gamblers, and saw and heard what they wanted to see and hear.
Everybody had looked for the discovery of another great bonanza
at the north end, and here it was. To be sure there was very
little of it in sight, but its great riches would be disclosed in time.
Experts and newspaper men vied with one another in spreading roseate reports of "crosscuts going forward in rich ore,"
when, in fact, they were standing still. There were rumors and

---

[6]Shuck's *History of the Bench and Bar*, p. 97 (1901).

[7]The boundary line between the two mines was in dispute, and the Sierra
Nevada claimed all of that ground. Litigation was averted by compromise
in December, whereby the Union shared in the ore bodies.

whisperings of rich disclosures known only to a few. The deal had the appearance of a widespread conspiracy to boost the stock, yet, after the public was first admitted, the mine was open to inspection most of the time. The workings were so simple and the facts so plain that no straight-thinking observer could be misled.

The Sierra Nevada had been extensively prospected for years without producing a ton of rich ore, but had always been a favorite gamble. At this time it was sinking a great third-line shaft, from which a long drift had been sent southward on the 1700-foot level to a point near the Union line, from which a winze was being sunk. In June of 1878 the stock began to advance mysteriously, and continued throughout July. The mine was then in the control of clever and successful Johnny Skae, who was quietly buying all of the stock he wanted before inviting the mine superintendents and newspaper men to inspect a new bonanza.

The winze had encountered low-grade ore at the depth of 2,000 feet, and this ore continued down almost to the 2100-foot level, meanwhile improving in value, some of it being very rich. At this point the ore passed beneath the winze, which continued on its course. A short crosscut on the 2100-foot level disclosed several feet of high-grade ore, from which A. E. Head took a piece of ore that assayed "nearly $900 to the ton." This was all the so-called experts and the newspaper men could see on August 30.[8] They were informed by Superintendent Charles Bonnemort that when the winze reached the 2200-foot level another crosscut would be driven back to the ore.

The "Chronicle" reporter thought it "a splendid prospect, with the strongest indications of proving a veritable bonanza," but added, "we must wait for that crosscut." Dan DeQuille, who had been prophesying another great bonanza at the North End, was most enthusiastic in his praise of the discovery from first to last. He saw what he wanted to see, like most of the others.

The stocks of the two mines had risen from day to day until toward the end of August, when they broke badly. The "Chronicle" of September 1 reported that President Skae had ordered a crosscut driven back to the ore at the 2150 level in order to confound the bears, who had driven the stock down from $86 to $45 a share. The newspaper went on to say that:

---

[8]*Virginia Evening Chronicle*, August 30, 1878.

The last eight or nine feet of the crosscut is in first-class ore; that in the face being the richest.   There work was stopped for the present.   * * *   Such men as Frank Osbiston, John Kelly, and Jim Rule declare that this is the finest face of ore they have ever seen on the Comstock.

This little crosscut later became known as "the coyote hole." The market was running wild, and Skae was taking no chance of spoiling his showing by extending the crosscut into perhaps barren rock.

Stocks immediately bounded higher than ever.   The excitement on the Comstock and in San Francisco rivaled that of the bonanza stock market at the end of 1874.   Sierra Nevada rose from $90 on September 4 to $170 on the 11th, to $200 on the 21st, and to $280 on the 27th.   Union followed along, reaching $182 on the same day.   Nearly all other Comstocks were booming; the less they were worth the higher they went in proportion.   Con. Virginia and California, the only two mines on the Lode paying dividends, rose very little.

It appears that Flood and Fair caught the infection and began to buy North End stocks, particularly in the Union and the Sierra Nevada.   Fair had resigned as superintendent of the Firm's mines in July, on the score of ill health, but came up from San Francisco on September 26 to pass on the new bonanza.   The winze had then reached the 2200-foot level, where a crosscut to the west passed through 20 feet of ore, some of it of excellent quality.

All that was to be seen in the mine at the time was the low-grade ore above the 2100-foot level; a little high-grade ore in a short crosscut on that level; nine feet of very good ore in a crosscut at the 2150-foot point, and 20 feet of ore in the crosscut on the 2200-foot level, only a portion of which could be called rich. The winze was on the boundary line between the Sierra Nevada and the Union, and the ore appeared to be pitching into the latter.

George T. Marye, Jr., tells the story of the deal at some length in the biography of his father,[9] in the course of which he recounts Fair's trip down the winze with Robert Sherwood, president of the Union.   Fair examined the winze and the crosscuts carefully, then turned to his companion and said:

Sherwood, you have 5,000 shares of Union haven't

---

[9]*From '49 to '83*, pp. 188–204 (1923).

you? Now I'll tell you what I'll do with you. I'll give you $200 a share for your lot. The stock is selling a little above $180 and I will give you $200 a share cash for your 5,000 shares if you want to sell.[10]

Sherwood hesitated a moment and agreed. In San Francisco the next morning Flood, on receipt of a telegram from Fair, sent $1,000,000 in U. S. Bonds to Sherwood's office and received the stock.[11] Marye says that Fair went to the office of his father in Virginia City and told of his bargain "possibly with the idea of strengthening the market"; but Marye, Sr., only shook his head and remarked that "he thought he could have got the stock for him cheaper." Flood had been buying Union quietly, and that 5,000 shares gave the Firm control of the mine. Fair left for San Francisco the following day.[12]

The "Enterprise" of October 18 noted the arrival of another member of the Firm: "John W. Mackay, of the Bonanza Mines, arrived here yesterday morning, direct from the Paris Exposition, to which he was a Commissioner." The meagerness of the showing in the winze astounded him, and he telegraphed Flood in code, "Fair is crazy."[13] Three days later he left for San Francisco with Dick Dey, who had come up to talk over business affairs.

September 27 was the top of the market: Sierra Nevada sold at $280, Union at $182. By October 3, Sierra Nevada had fallen to $242;[14] a week later it was quoted at $195, and on October 22 at $165. Fair, who knew when to duck an unpleasant situation, left for the East on October 19.[15] The market held its own, and even rose a little until November 18 when a panic seized the

---

[10]Id., p. 198; *History of San Francisco Stock Exchange*, p. 277 (1910).

[11]George L. Upshur. James E. Walsh said that Flood had great faith in Fair's judgment of a mine. That transaction evened Sherwood's score with Flood, who bought 1,000 shares of Con. Virginia from Sherwood in November 1874 for $100 a share and made a large profit.

[12]"Col. J. G. Fair and several other magnates left last night for San Francisco." *Gold Hill News*, September 28, 1878.

Postmaster General Key and party arrived at Virginia City on October 13, 1878, and were welcomed by Alderman James Orndorff. (*Daily Territorial Enterprise*, October 14, 1878).

"Jim Orndorff was the most prominent and successful saloonkeeper in Virginia City for years. Lived like a prince and kept a pack of hounds to hunt jack rabbits and coyotes, and deer in the Sierras in the fall.

[13]James E. Walsh.

[14]*Gold Hill News* of October 2d reports that the Sierra Nevada winze is going down on the richest ore ever discovered on the Comstock.

[15]*Daily Territorial Enterprise*, October 20, 1878.

GENERAL GRANT AND PARTY

(See bottom opposite page)

speculators. Sierra Nevada opened at $200, broke to $90 the next morning, and to $65 on the 20th. Never before in the history of the Exchange had a leader lost half of its value in one day. The bottom had dropped out of the market. Nearly all of the stocks had been bought on margin account and the lenders could not sell fast enough to protect themselves. Some that had been smart enough to sell at a profit could not resist coming back into the market when stocks continued to rise, and lost with the rest.

> This break, of course, carried with it the stock of every mine on the Comstock, depreciating the whole to the extent of tens of millions of dollars. * * * Thousands on the Coast, both the rich and the poor, have seen their all swept to utter destruction.[16]

The "Deal" profited few men, and only Skae greatly. It is said that he lost his fortune a few years later.

GENERAL GRANT AND PRESIDENT HAYES VISIT THE COMSTOCK

Among the many distinguished men who visited the Comstock in the '70s and '80s were Ex-President Grant and President Rutherford B. Hayes. General Grant and party, consisting of General and Mrs. Grant and their son U. S. Grant, Jr., and wife, who were returning from a tour of the world, lasting two years, arrived on the Comstock on October 27, 1879, and remained for three days.

Never before was such a reception given to any one in Nevada. It was a wholehearted western welcome to a popular idol. Business was practically suspended; parades, banquets, speeches, bonfires, and entertainments followed in quick succession. The town was decorated as for a Fourth of July. Union and Confederate veterans marched in the parades. General Grant's reply to the Mayor's speech of welcome was "I thank you." The party visited some of the leading mines, and were taken down the C. & C. shaft for a trip through the Con. Virginia and the California lower

---

[16]*Gold Hill News*, November 20, 1878.

During the excitement it was thought that the winze was going down on a vein of rich ore, which acted rather peculiarly, but the fact developed that the winze had crossed two small vertical lenses of ore which stood one below the other. The production of the two little ore bodies was only $2,000,000.

GENERAL GRANT AND PARTY—Standing at the mouth of "C. & C." Shaft, dressed to go down into the Consolidated Virginia and California mines, October 28, 1879. Members of party, left to right: Mr. J. W. Mackay, Mrs. M. G. Gillette, U. S. Grant, Jr., Mrs. U. S. Grant, Gen. U. S. Grant, Mrs. J. G. Fair, Gov. J. H. Kinkead, Col. J. G. Fair. S. Yamada

levels.   The heat in places ranged as high as 130 degrees, but the General wanted to see everything.   The ladies and some others of the party were satisfied to return after reaching the 1700-foot level.[17]   At the Con. Virginia assay office they saw "about twenty tons" of silver and gold bullion stacked up in bars like cordwood. Adolph Sutro entertained the party at his mansion, after which they returned to Virginia City by way of the tunnel and the C. & C. shaft.   Fair, who was in poor health, came up from San Francisco to be of the party.

John Russell Young there met Mackay for the first time.   General Grant said in introducing him:

> I want you to meet a man whom I am proud to know and call my friend.   He is a man of great ability and some day will make his mark outside of the mines.   He would make a great general had he been trained for the army.[18]

President and Mrs. Hayes, accompanied by Major General William T. Sherman, Major General Alexander McDowell McCook, and Secretary of War Alexander Ramsey, arrived on the Comstock September 7, 1880.   They were welcomed at Gold Hill by Governor Kinkead and local dignitaries, and escorted to Virginia City by a great parade, with bands playing, flags flying, and mine whistles blowing.   The Comstock felt highly honored by a visit from the President and those distinguished generals.   Enthusiastic crowds gathered in front of the International Hotel and called for the President and the Generals, all of whom made brief and happy speeches.   After the banquet the members of the party were taken through the Con. Virginia and California mines. They remained but one day.

President Hayes was greeted by the following distinguished reception committee: Governor J. H. Kinkead, Lieutenant Governor Jewett W. Adams, John W. Mackay, R. H. Taylor, C. H. Belknap, D. O. Adkison, Richard Rising, J. C. Curry, L. T. Fox,

---

[17]The *Daily Territorial Enterprise* of the 29th says that the following correspondents accompanying the party made the trip through the mines: T. B. H. Stenhouse of the *New York Herald;* J. W. Robbins of the *Chicago Inter-Ocean;* and C. R. Brodix of the *Bloomington (Ill.) Leader.*

[18]*New York Herald*, July 21, 1902.   We may believe that the General's words were fewer.   The extract is from Young's *Men and Memories.*

Young was a member of the party and wrote a book, *Around the World With General Grant.* He became friendly with Mackay, but was no more successful than others in obtaining biographical material for his sketch in *Men and Memories.*

C. J. Hillyer, R. P. Keating, W. H. Patton, L. P. Drexler, J. Minor Taylor, B. C. Whitman, George T. Marye, John Kelly, Matt. Canavan, T. G. Taylor, J. H. Graham, Alf Doten, J. P. Smith, J. C. Hampton, W. G. Hyde, I. L. Requa, A. B. Elliott, Thomas Gallagher, A. F. McKay, W. Hy Doane, J. S. Young, F. H. Hart, M. J. Burke, Bishop O. W. Whitaker, George Brown, Fred Schroeder, Charles Mueller, Spiro Vucovich, G. E. Caukin, J. H. Harris, W. S. Bronson, J. B. Overton, John Piper, William Sharon, William Pennison, E. B. Harris, Thomas Rooney, M. Kennedy, Joseph Price, Andrew Fraser, E. Strother, J. J. Sheppard, D. L. Brown, B. Dellepiane, George W. Birdsall, William Price, Jr., W. P. Pratt, A. Ash, William Wright "Dan DeQuille," M. Banner, Dan Lyons, Robert Patterson, Wells Drury, W. D. C. Gibson, B. F. Hazeltine, Sam L. Jones, J. R. Stuart.[19]

END OF BONANZA PERIOD

The glory of the Comstock passed with the year 1880, and with it a life as high-spirited, colorful, and romantic as that in "The Days of '49." Not only was it finished as a profitable mining region, but most of the ambitious, energetic men had departed or were soon to leave—the mining men to the ends of the earth, where they made history.[20] Many of those men became prominent and successful in their new fields; among them Marcus Daly and David Keith, who had been Comstock mine foremen. Daly was the founder and manager of the great Anaconda copper mine at Butte, Montana, where many old Comstockers found employment. Keith was one of the organizers and principal owners of the rich Silver King silver-lead mine at Park City, Utah. The lawyers scattered throughout the West, the larger number to San Francisco, where a number of them soon took rank among

---

[19]*Daily Territorial Enterprise*, September 5, 1880.

In no other place in the world would the names of leading gamblers and saloonkeepers appear on such a committee, which included Bishop Whitaker; yet there is Joseph R. "Joe" Stuart, who kept the finest faro rooms on the Comstock for years; Robert "Bob" Patterson, proprietor of the International Saloon and faro rooms; and W. D. C. Gibson (William De Witt Clinton Gibson, in full, familiarly known as "Bill"), the proprietor of a similar establishment at Gold Hill. Manhood and brains were the only tests on the Comstock.

[20]Dan DeQuille wrote in 1889: "Men who were graduated on the Comstock are now to be found in all parts of the world. They early went to Idaho, Montana, Utah, Colorado, New Mexico, Arizona, Alaska, and British Columbia. Old Comstock foremen and superintendents are today in charge of mines in Mexico, Central America, South America, Australia, Africa, China, Japan, and all other regions where there is mining for the precious metals." *A History of the Comstock Mines*, p. 45 (1889).

the leaders. Business men and others of various vocations rees-
tablished themselves elsewhere as best they could, but it was
a sad change for many of them, especially those of middle age.
Instead of the old free life among congenial spirits they found
themselves comparative strangers in new surroundings and almost
unwanted.

The Con. Virginia paid its last small dividend in 1880 (until
resumed during the low-grade period from 1886 to 1895). The
stock market was all but dead. In 1881 the production of the
Comstock was only $1,075,600, and that from unprofitable ore.
Thereafter nothing remained for the leading mines but to carry
on intensive deep mining, which gradually ceased. Then followed
the low-grade period wherein the mines returned to the old upper
levels to extract the ore remaining in the margins of the old
bonanzas. The decreased population was largely foreign, but the
remaining old-timers kept alive the spirit of earlier days as best
they could.

From 1859 to 1882 the Comstock mines produced $320,000,000
from ore and tailings, and paid $147,000,000 in dividends (includ-
ing private milling profits). Meantime the assessments levied
and the expenditures by private companies and individuals
amounted to not less than $92,000,000, leaving a net profit of
$55,000,000, nearly all of which accrued in the '70s.[21]

Bonanzas must have an end, but that one made a record in
mining history, with a production of $105,014,498 in seven years
and the distribution of $74,250,000 in dividends. The Con. Vir-
ginia paid dividends of $1,080,000 a month for thirty-four months
after it got well started, and a total of $42,930,000. The Cali-
fornia distributed an equal amount for twenty-six months and a
total of $31,320,000. After the rich ore was exhausted dividends
decreased rapidly. The Con. Virginia paid its last in 1880, the
California in 1879.[22]

In 1875 the Con. Virginia produced 169,049 tons of ore yielding
$16,076,680, or $95 a ton, from which $11,448,000 was distributed
in dividends. In 1876 the mine produced 145,466 tons, yielding
$15,315,614, or $105.31 a ton, out of which $12,960,000 was paid
in dividends. The California made its highest production in 1877,

---

[21]A table giving the production of the various mines from 1859 to 1882,
together with tonnage, value, dividends, etc., will be found in the Appendix.

[22]The two mines paid additional dividends amounting to $3,898,800 during
the low-grade period from 1886 to 1895. The production statements will be
found in the Appendix.

when 213,683 tons yielded $17,879,948, or $83.20 a ton, and dividends amounting to $14,040,000 were paid.[23]

Not only was the bonanza period at an end in 1880, but the Comstock was finished as a profitable mining region. Thereafter the few dividends paid were swamped by an appalling roll of assessments. The total production of the Lode in 1881 was only $1,075,620, the lowest in any year since 1860. Stocks had fallen in the same measure. The market value of all of the mines on the Lode was less than $7,000,000—a decline of $293,000,000 in six years. The only hope left was in deep mining; the Lode had never failed them before and men believed that other great bonanzas would be found below.

---

[23]All statements of the production of the two bonanza mines are based on the coin, or market, value of silver after it was demonetized in 1873. The annual reports, in accordance with Comstock practice, continued to value the silver produced at $1.2929 per ounce, and charged the discount as an expense of operation.

CHAPTER XXIII

Life on the Comstock in the '70s—Comstock Millionaires—Notable Comstock Mine Superintendents.

Life on the Comstock had always been full of interest and enjoyment, but the '70s overtopped all that had gone before. People were more comfortably housed, a bountiful supply of water had been brought in from the tops of the Sierras, the streets were macadamized with refuse from the old mine dumps and lighted with gas, the city had railroad connections both east and west, people traveled far more than in earlier years, social life took on wider aspects, the Opera House presented a constant stream of dramatic productions, and entertainments of all kinds multiplied.

A correspondent wrote of the city in 1878:

> There is no place in the world of the population of Virginia City possessing so many interesting and peculiar characteristics. It has all the features of a great metropolis, hotels, stores, places of amusement, churches, clubs, banks, four or five daily journals, foundries, machine shops, a railroad, water works, furnishing an abundant supply of pure soft water from the Sierras, an active and efficient paid fire department, etc.[1]

The years since the feverish and riotous boom of 1863 had left their impress—the Comstock was more settled in its ways, more metropolitan. In outlook it had changed little. Although the foreign element largely exceeded in numbers, the pioneers remained in control and their spirit dominated everything. They were years older yet scarcely less youthful in their enthusiasms. Everybody continued to take part in the endless celebrations and social affairs. Life was still a great adventure. Of course, all were going to get rich and retire to California, but, in the meantime, they lived for the day. Humor was still the prevailing spirit, although lacking something of the freshness and spontaneity of earlier years. A more self-reliant, independent, brave, and generous community of men and women did not exist. The world was young and they thought themselves fortunate to be living at such a time.

---

[1] *Mining and Scientific Press*, September 12, 1878.

There were 2,200 buildings in Virginia City in 1880, of which 92 were of brick. Lord's *Comstock Mining and Miners*, p. 352.

"If absence from care is happiness," wrote the editor of the "Virginia Evening Chronicle," "the population of Virginia City ought to be the happiest in the world. There is a feeling of independence here scarcely ever experienced elsewhere, and a freedom from the artificial trammels of society which in older communities are based on wealth, birth, and position. In other words, every man here is a man in his boots. * * * It is hard to find anywhere a more hopeful, sanguine, and independent population than that of Virginia City."

The Comstock, far removed from the great outer world, was a little world to itself, composed of two almost distinct elements: The one, made up of the business and saloon life on "C" Street, the popular red-light district on "D" Street, and the gambling, opium-smoking Chinese settlement below; the other, and by far the larger part of the community, consisting of the thousand families living chiefly above "C" Street, of whom nothing is said by the chroniclers.

The majority lived simple lives compared with the present hurried, fretful existence, and were more content. Artificial wants have multiplied and happiness is farther off than ever. Our delight in simple things would be thought childish nowadays. A stroll on the long summer evenings, much neighborhood visiting, and little quiet parties, an occasional evening at the theater to watch a play or hear a lecture, or perhaps to a concert given by local talent, for everybody sang in those days. The attractions at the theater ranged from Shakespeare to minstrel shows and melodrama, but there were no dirty sex plays and no naked women on the stage, which modern taste appears to demand.[2]

It was a city of schools and churches, of which we read little in the books on the Comstock. School entertainments were always well attended. The churches were social centers and their "socials" and "fairs" contributed largely to their finances. The annual picnics of various organizations to Bowers' Mansion drew trainloads of people, including many children.

What boy or girl ever forgot the first picnic at Bowers' Mansion

---

[2]"The Black Crook," which arrived in 1867, was the first leg show to appear on the Comstock and took the town by storm, although a tame affair compared with present-day disclosures. The respectable element was shocked. In the late '70s there was an irruption of British Blondes and Red Stocking Blondes, whose abundant fleshly charms, encased in tights, were displayed on flaming posters.

Wells Drury, on pages 54 to 61 of *An Editor On The Comstock Lode* (1936), gives some enjoyable reminiscences of the dramatic offerings in the '70s and '80s.

in Washoe Valley, where for the first time in their lives they saw clear running brooks, great pine trees, wide meadows spangled with flowers, and, beyond the meadows the shimmering expanse of Washoe Lake? It was a trip to Paradise.

Winter, with its never-ending balls and parties and other entertainments, was even more joyous than the summer season. Coasting and sleigh-riding brought forth merry crowds day and night.

The spirit of helpfulness, which was the chief charm of the camp, grew out of mutual dependence in earlier more primitive days, and had become spontaneous. There was no thought of reward—the pleasure was in giving. Liberality in money matters was almost a vice, or seemed so in later years of deprivation. Liberality of thought was equally characteristic of the people. To this day you can tell an old-time Comstocker by his breadth of view. In later years when two of them met they came together as brothers, with thoughts in common and memories that brought tears to their eyes.

The Comstock continued to be Mid-Victorian in dress and manners. To be a gentleman or a lady was the ideal. The good women were held in highest esteem, although more prominent in the life of the community than in earlier years. They even took a hand in politics when striving to bring about social reforms. Upon their insistence State laws were enacted prohibiting minors in saloons, and requiring gambling to be carried on behind closed doors. They labored for years to make gambling unlawful, without success.

The men foregathered in the clubs and in the halls of the many fraternal and other organizations. Every military and volunteer fire company had its headquarters, the German their Turn Verien, and the Miners' Union its own hall and library. The various races had their gathering places.

But it was the saloons that contributed most to the good-fellowship of the camp. Nothing puts men upon a friendly footing more quickly than to drink together. Views were exchanged and not a little important business transacted at the bar. The reverse of that picture is the fact that heavy drinking was the curse of the Comstock. Some men drank as much as a quart of hard liquor a day, in many small drinks, and carried it off for years. The saloons were of all grades, like the population; from first-class places where the price of a drink or a cigar was a quarter, to the lowest dives. As a rule the various elements sought their own kind. Virginia City was the "good-time town" of the region. Men came from miles around after payday and contributed not a little to the sporting life—to which the many visitors from San Francisco added their share.

The stock market was the nerve center of the region, and the reports from the San Francisco exchanges, which came morning and afternoon, invariably gathered crowds in front of the brokers' offices. Nearly everybody had stocks and all wanted to know how the market was going. Copies of the reports were posted at the heads of the shafts in the principal mines for the information of men coming off shift. Not infrequently a miner, noting a big rise in his stocks, would throw his lunch bucket in the air, exulting "To hell with work!" But it would not be long before he was back asking for a job.

Whether stocks were going up or down life went on at the same high pitch. A panic or a great fire dampened their spirits only momentarily. Life was a gamble—better luck next time. Even our lordly mine superintendents, who were subject to coded instructions from San Francisco, guessed the market wrong quite as often as well-informed outsiders.

Eveybody wanted to get rich—always out of stocks or a mine, as other successful men had done. A saving or an acquisitive man was almost unknown. They spent money like leaves and they owned a forest. The theory behind getting rich was that you would have more to spend. Competition was not ruthless. No one wanted to break his neighbor. "Live and let live" was the prevailing spirit. There was no desire to manage others. Time was not money. There was always time to do something for somebody else. Time to talk and exchange views about everything under the sun; time to play, for they were a pleasure-loving people. Marye, who knew them all, wrote that "the commanding spirits were moved by a big-hearted generous desire for achievement."[3]

This writer has an abiding memory that men took pride in keeping their word, and that a liar was despised. Another vivid recollection is that the leading men and women, the pioneers and their wives, always dressed well and bore themselves with an air of quiet dignity. The "grand manner" seemed natural to them. An old resident of San Francisco, a bank president, remarked to this writer recently, "You could always tell one of those old-time Nevadans; they wore such good clothes and carried themselves with such an air."

---

[3] From *'49 to '83 in California and Nevada*, p. 62 (1923).

The Comstock lived well. One looks back upon the '70s with some surprise at the time and thought given to food. Those who could afford it, and especially the men about town, lived on the fat of the land. The markets were as well supplied as those in San Francisco, and the restaurants equally as good. Thick, juicy steaks and roast beef headed the list in popularity; men often ate steaks for breakfast. Oysters, Eastern and California, were almost a staple. Fish and game were abundant; trout from the Truckee and Lake Tahoe, caught by the Indians.

All kinds of fish came from the Coast. Sagehens and ducks and grouse were Nevada products in season. California provided the quail, chickens, and turkeys. The restaurants often featured a large live sea turtle. Fruits of all kinds from California were items of daily fare, as well as an abundance of fresh vegetables. There was hardly any kind of food that could not be obtained, even to foreign delicacies. The bakeries were first class; only a few of the housewives baked their bread. Milk from the insanitary dairies perhaps helped to spread typhoid fever.

The majority of the people with families lived well, but simply. Unmarried miners and other single men boarded at the many restaurants, which provided excellent fare at $30 to $35 a month. The cost of living was less than it is today; some of the items were higher, others lower. The smallest piece of money was a silver dime, ten cents, which was the price of a small basket of strawberries in season. The only money in circulation was gold and silver, although in the East they had nothing but paper currency until after the resumption of specie payments in 1879.

Frontier life teaches men to think for themselves. This was especially true of early mining camp life in California and Nevada. The philosophy of life of the pioneers was largely the result of observation and experience. There was little dependence upon Divine Providence—men had learned to rely upon themselves. Their religion was a sort of golden rule. They encouraged and supported churches because of their influence upon the community. Ingersoll expressed their views about the Bible, and the Darwinian theory appealed to their intelligence. The mechanistic theory of the universe appeared to them as the only reasonable explanation. Man himself was only a machine, born to function according to his gifts; some fearfully and wonderfully made; others mere automatons, with many gradations between. The Comstock was a laboratory of life, in which it appeared that all men are born unequal. All of the races of the world were there and all types of men and women, from the highest to the lowest. One learned more of the human animal in a few years in that congested community than could be acquired in a lifetime in a conventional town.

The average of intelligence was higher in the early '60s than in the '70s when the foreign-born came in increasing numbers, although the highly intelligent members of the community maintained their standards; they read the best literature, were exceedingly well informed on all of the topics of the times, and were engaging conversationalists.

Dick Dey, in his heartfelt obituary of Mackay,[4] said: "There never was another like him, even in the big days when men seemed to have a chance to be bigger than they are nowadays." This writer has often wondered why the outstanding men of those times seemed "bigger" than those of today. That was equally true in San Francisco. The explanation appears to be that they were the product of pioneer conditions, which brought out the best as well as the worst. Men are made by struggle.

COMSTOCK MILLIONAIRES

Comstock millionaires were far fewer than is generally supposed, and all were created in the '70s. In the early '60s the Ophir and Gould & Curry bonanazs did not create a single millionaire. The fortunes of George Hearst, John O. Earl, Robert Morrow, A. E. Head, Andrew B. McCreery, and Charles N. Felton had their beginnings in the Gould & Curry and the Savage, although none of them acquired a million dollars there.

The Crown Point-Belcher bonanza brought millions to John P. Jones, Alvinza Hayward, William Sharon, William C. Ralston, and D. O. Mills.

The Con. Virginia bonanza created a longer list: John W. Mackay, James C. Flood, James G. Fair, William S. O'Brien, General Thomas J. Williams, David Bixler, Robert N. Graves, and Edward Barron.

E. J. "Lucky" Baldwin sold his stock in the Ophir to Sharon in November 1874 for $2,500,000. Robert Sherwood and Johnny Skae got rich during the "Sierra Nevada Deal." Sutro's wealth came from the sale of his stock in the Tunnel. Archie Borland and William M. Lent were miners and speculators. William S. Hobart's large fortune came chiefly from lumbering, and incidently from mines and mills.

---

[4]*San Francisco Examiner*, July 21, 1902.

Nearly all of those millionaires retained their wealth, in part at least. During stock excitements prospective millionaires in San Francisco and on the Comstock were as plentiful as blackberries, but the inevitable decline left them as poor as winter.

Isaac L. Requa acquired a substantial fortune after the rich Belvidere ore body was discovered in the Chollar-Potosi in 1869. "Sandy" Bowers' profits did not exceed half a million, and he spent it as fast as it came. When he died in 1868 his entire estate was appraised at $88,998.

The San Francisco stock brokers were more uniformly successful than any other class connected with the Comstock mines. More than fifty of them made substantial fortunes, and it is said that the following acquired a million or more: James R. Keene, John D. Fry, Col. E. E. Eyre, John W. Coleman, George T. Marye, Sr., B. F. Sherwood, James Latham, and C. W. Bonynge.

NOTABLE COMSTOCK MINE SUPERINTENDENTS
1859–1871 AND 1871–1886

1859–1871

Captain William L. Dall—Ophir.
Philipp Deidesheimer—Ophir and others.
W. W. Palmer—Ophir.
H. H. Day—Ophir.
A. E. "Hog" Davis—Ophir.
James G. Fair—Ophir; Hale & Norcross.
Robert Morrow—Savage; White & Murphy.
Charles L. Strong—Gould & Curry.
Pat McKay—Potosi.
Charles Bonner—Gould & Curry; Savage.
Louis Janin, Jr.—Gould & Curry mill.
I. Adams—Chollar.
Capt. Sam T. Curtis—Meredith; Con. Virginia; Savage, and
    others.
C. C. Thomas—Uncle Sam; Hale & Norcross.
J. M. Walker—Bullion.
R. K. Colcord—Imperial-Empire Shaft.
John W. Mackay—Caledonia Tunnel; Milton; Bullion.
John D. Winters—Yellow Jacket; Kentuck.
C. C. Stevenson—Various Gold Hill mines.
P. S. Buckminster—Imperial.
John Lambert—Belcher and others.

Robert N. Graves—Empire.

Enoch Strother—Baltimore.

Capt. Thomas G. Taylor—Yellow Jacket and Best & Belcher.

Seth Cook—Sierra Nevada.

Gen. C. C. Batterman—Crown Point; Gould & Curry, and others.

Harvey Beckwith—Mexican; Chollar-Potosi.

Charles Forman—Eclipse and others.

James Rule—Hale & Norcross and others.

J. P. Jones—Kentuck; Crown Point.

W. E. Bidleman—Utah.

J. D. Greentree—Imperial.

I. L. Requa—Gould & Curry; Chollar-Potosi, and Sharon's mills.

D. B. Lyman—Mexican and other mills.

Captain Lloyd Rawlings—Savage mills.

James P. Woodbury—Various mills.

H. G. Blasdel.

D. E. Avery—New York & Washoe.

Robert Apple—Minerva.

L. U. Colbath—Challenge.

R. P. Keating—Overman and others.

1871–1886

William E. Sharon—Gold Hill mines.

William Skyrme—Hale & Norcross, etc.

John Lambert—Sierra Nevada and others.

H. G. Blasdel—Rock Island and others.

Frederick Thayer—Julia; Ward.

Chas. M. Bonnemort—Sierra Nevada.

D. H. Jackson—North Con. Virginia.

Albert Lackay—Trojan and others.

Archie Borland—Empire.

Jas. P. Woodbury—Mariposa and other mills.

J. P. Jones—Kentuck; Crown Point.

Sam L. Jones—Kentuck; Crown Point.

W. H. "Hank" Smith—Belcher; Overman; Caledonia, and others.

Gen. C. C. Batterman—Imperial and others.

Capt. Thos. G. Taylor—Yellow Jacket.

Frank F. Osbiston—Baltimore; Savage; Gould & Curry.

Charles Forman—Overman; Caledonia; Forman Shaft.

James G. Fair—Hale & Norcross, and Bonanza Firm's mines.

Capt. Lloyd Rawlings—Rawlings Shaft.

I. L. Requa—Chollar-Potosi; Combination Shaft; Sharon's mills.

Sam T. Curtis—Ophir; Union and Mexican; Justice, and others.

W. T. "Joggles" Wright—Sierra Nevada.

I. E. James—Yellow Jacket.

Pat Kerwin—Gould & Curry.

A. C. Hamilton—Savage and others.

Philipp Deidesheimer—Ophir; Hale & Norcross; Justice, and others.

William Hardy—Ophir; Imperial.

W. H. Patton—Bonanza Firm's mines, 1878–1887.

D. B. Lyman—Mexican Mill; Bonanza Firm's mills, 1875–1897; Bonanza mines, 1887–1897.

H. H. Penoyer—Gould & Curry.

R. P. Keating—Hale & Norcross and others.

Ed. Boyle—Bullion; Alta.

T. F. Smith—Con. Virginia; Justice.

Charles Derby—Alta.

W. B. Sheppard—Utah.

John Egan—Andes and others.

James Rule—Gould & Curry; Utah and others.

Milton G. Gillette—Savage.

Matt Canavan—New York.

M. C. Hillyer—Silver Hill.

C. C. Stevenson—Stevenson and others.

Alex McKenzie—Sierra Nevada.

Enoch Strother—Baltimore.

J. M. Walker—Dayton.

C. C. Thomas—Sutro Tunnel.

CHAPTER XXIV

Miners' Wages and Hours — Heat and Ventilation — Giant Powder, Burleigh and Diamond Drills—Lumber and Firewood—The V Flume.

Comstock miners were the lords of labor and gloried in it. They were the pick of the world; their wages the highest, their hours the shortest; they were men among men. Independent-minded, like the rest of the community, they resisted all attempts to reduce wages below $4 a day, and first organized for that purpose as early as 1863. During the hard times of 1865 and 1866 they submitted to a reduction by some of the mines, but were quick to unite again in 1867 with only partial success until 1872.

While in 1864 and again in 1867 the miners organized and marched in bodies to assert their demands, there was no violence nor any destruction of property. John Trembath, the Cornish foreman at the Uncle Sam mine, who was bound to the hoisting cable and jerked up and down, might question the statement that there was no violence.[1]

Shinn says[2] that "On one occasion a superintendent (Charles Bonner of the Gould & Curry) who had attempted to cut wages, was concealed in the home of a priest (Father Manogue) or he would have been torn limb from limb by the indignant miners." An overstatement, no doubt.

Ten hours was a shift during all of the early years, but, as conditions underground became more intolerable, the hours of men working in such places were reduced to eight. In 1867 the constitution of the newly formed Miners' Union provided that all men working underground should receive $4 for an eight-hour shift. That rule was not enforced, it appears, but became uniform after John P. Jones, then candidate for the U. S. Senate, ordered that on and after April 1, 1872, the eight-hour day should apply to all men working underground in the Crown Point mine.[3] The other mines, most of which were controlled by Sharon, who was also a candidate, quickly adopted the same rule, which thereafter prevailed on the Comstock.

[1]Lord's *Comstock Mining and Miners*, pp. 183–190, 266–268.
[2]*The Story of the Mine*, p. 250 (1896).
[3]*Daily Territorial Enterprise*, April 2, 1872.

A GROUP OF COMSTOCK MINERS

The Miners' Union grew into a benevolent institution.[4] **The** mines would not permit any man to work underground who **was** not a member, and deducted the monthly dues of $2 from **each** man's pay.[5] In return, the Union cared for sick and **disabled** miners, although the companies usually contributed. It was **the** custom when a man was killed for each miner to contribute **a** day's pay to the family.

After the first few years the number of foreign-born **miners** increased steadily. Lord (pp. 383, 384) gives statistics for **the** year 1880: 1,996 miners, of whom 394 were American, 691 **Irish,** 543 English, 132 Canadians, and the rest from everywhere. **The** average age was 35, average weight 165 pounds, average **height** 5 feet 9 inches. The majority were married.

Shinn overstates the standard of living enjoyed by the miners: "Every observer of the Comstock in its palmy days noted **the** universally high standard of living. Not only the necessaries, but the luxuries of life formed the daily fare of the miners." That is true of the unmarried miners who boarded at first-class restaurants, but this writer well remembers many hundreds **of** small homes in which there did not appear to be room for **the** large families of children. They lived well but simply.

Death lurked everywhere in those mines, but there was no **fear,** only constant watchfulness, coupled with an element of fatalism. At times one man was killed or fatally injured every week, **and** one more or less seriously injured every day. Accidents **were** deemed risks of the employment, and it does not appear that **the** companies were sued for damages.

The shafts were among the most dangerous places in the mines, notwithstanding skillful timbering and the perfection of hoisting machinery. Individual deaths from falling and otherwise were not infrequent, and more men were killed on the cages at one time than in any other accidents below. Three of the worst occurred within one year.

On December 2, 1879, when 17 men were being hoisted in the Union Shaft, the engineer pulled the wrong lever, and the **cage**

---

[4]The Miners' Union had its own building and the largest general library in the State, 2,000 volumes.

[5]There was no ceremony attached to joining the Union. When this writer went to work in the mines he merely enrolled at the Secretary's office and received his card. On payday his dues were deducted from the check.

The average wage paid to common miners in California in 1874 was $1.50 to $2 per day, says the *Mining and Scientific Press* of February 2, 1875, which prints a list of all sorts of employees. "No wonder the Comstock miners were the pick of the world."

and the skip containing the men, instead of stopping at the collar of the shaft shot with lightning speed into the sheaves at the top of the 40-foot gallows frame, crushing the cage and spilling the men all over the floor of the hoist house.   Two were killed outright and seven permanently injured.   The escape of the others from death was little short of miraculous.   Luckily none fell down the shaft.

On June 18, 1880, as eight men were standing on a skip waiting to be hoisted from the 2800-foot level of the New Yellow Jacket shaft, a car loaded with steel drills, which was being hoisted in an adjoining compartment, caught an obstruction when near the top of the shaft (probably a drill in the car got loose), and the contents of the car were spilled upon the men standing half a mile below.   Five were killed instantly, one seriously hurt, and two injured slightly.

Nine men met instant death three months later in the Imperial shaft.   They were coming up on a cage when the steel hoisting cable broke.   The weight of the heavy cable which piled on top of the cage proved too much for the safety clutches and the cage and men dropped to the bottom.[6]

HEAT AND VENTILATION

The problem of ventilating the mine, that men might live and work, became as important as the extraction of ore.   Miles of drifts, crosscuts and raises were driven for no other purpose. The mines were connnected on many of their lower levels chiefly to promote the circulation of air.   The main shafts became the chief means of ventilation.   An automatic circulation was created by the fact that some of them stood at higher elevations than others.   Ten of the upper shafts were used as "upcasts," drawing the steaming fetid air from below, while six shafts, standing on a lower line, became "downcasts," carrying great volumes of fresh air to the lower levels.   Clouds of steam rose constantly from the mouths of upcast shafts.   Doors were placed at various points underground to regulate air currents.   Revolving fans, called blowers, were installed in many places, driving the air forward or sucking it out, while air compressors supplied remote workings.

Exposure to sharp changes in temperature was another danger

[6]Lord has a depressing chapter on *The Pains and Perils of Mining*, pp. 389–406 (1883).

The *Daily Territorial Enterprise* of February 24, 1877, has an illuminating article on *The Hazards of Mining on the Comstock*.

when men were hoisted from stifling levels to perhaps a snow-storm on the surface.

Miners commonly worked in temperatures ranging from 100 to 125 degrees, but observers agreed that owing to their superiority they accomplished as much as men in other camps working under normal conditions.[7]

No other mines in the world have encountered such heat and such floods of scalding water. The highest temperature of any considerable quantity of water (170 degrees) was recorded by the flood on the 3,000-foot level of the New Yellow Jacket shaft in November 1880. That water was first struck at a depth of 3,080 feet by a drill hole from the bottom of the New Yellow Jacket shaft.[8] Soon afterward the pump rod broke. The sudden jerk caused the huge cast-iron fly wheel to fly to pieces, and pumping was not resumed for six months. Meantime the Gold Hill mines were flooded.

Superintendent Taylor of the Yellow Jacket said in his inter-esting diary that the temperature of the rock in a dry drill hole on the 3,000-foot level recorded 167 degrees—"using cold water from the surface to spray the rock." The heat on the lower levels, he said, gave men cramps in the stomach.

The flood from the 2800-foot level of the Exchequer mine, which again drowned the pumps at the New Yellow Jacket shaft on February 13, 1882, averaged 157 degrees.[9] Again the Gold Hill mines were flooded. When they failed to agree to contribute to the pumping expense of the New Yellow Jacket that great plant was closed down, and the water in Gold Hill slowly rose to the Sutro Tunnel level. Water at 150 to 167 degrees tempera-tures will cook food, and men died from a brief submergence.[10]

GIANT POWDER, BURLEIGH DRILLS, DIAMOND DRILLS

Black powder was the only explosive used in the mines until 1868, when the Gould & Curry experimented with "giant powder"

---

[7]Prof. John A. Church (author of *The Comstock Lode, 1879*) in *Mining and Scientific Press*, August 9, 1879; Becker's *Geology of The Comstock Lode*, p. 4 (1882).

[8]Becker's *Geology of the Comstock Lode*, p. 230 (1882). Becker speaks of "The immense pressure which the water often shows on being tapped by the drills in the lower levels."

[9]Annual report of Superintendent Thomas G. Taylor, July 1, 1882.

[10]Becker says: "By far the greatest obstacle to mining on the Comstock has been the heat, which increases about 3 Fahrenheit for every additional hun-dred feet sunk, and which seems likely eventually to put an end to further sinking."

(dynamite), which had been invented by Alfred B. Nobel, that great Swedish chemist and engineer, in 1863 and was patented by him in 1867.[11]   In the latter year the Giant Powder Company began to manufacture it in California under the name giant powder, and so it was known in the West for many years.[12]   The test at the Gould & Curry was unsatisfactory, but within two years the explosive was in common use.   The miners at Grass Valley and Nevada City, California, organized a general strike against the use of the powder on the ground that it was highly dangerous and that the fumes made them sick.   Many of them came over to the Comstock only to find that they had to work with the detested stuff.

The Burleigh drilling machine was first installed in the Yellow Jacket in 1872,[13] and shortly thereafter all of the larger mines were equipped.   The drill was not only cheaper and more effective than hand work, but the compressed air employed in running it proved of inestimable value.   Deep mining would not have been possible without it.   Compressed air under high pressure was carried to the remotest workings, not only to run the drills but to furnish power for small hoists and pumps and ventilating blowers and to forward fresh air to the miners.   Lord, p. 365, is in error in stating that the serviceability of machine drills was first demonstrated in the Sutro Tunnel in 1874.

Diamond drills were first used in the Sutro Tunnel in April 1872, and soon came into general use for prospecting unexplored ground.[14]   Unfortunately no valuable ore was discovered by that means, but the drill became indispensable for boring ahead of advancing workings to learn whether a large volume of hot water

---

[11]The world remembers him chiefly as the founder of the Nobel Prizes.

[12]*U. S. Mineral Resources* for 1867, p. 97 ; Id. for 1869, pp. 55, 489–496.

On June 29, 1873, at 11 o'clock p. m., the McLaughlin & Root building on the corner of Taylor and B Streets, was blown to pieces by the explosion of 100 pounds of Hercules powder, 6 cases of nitro-glycerine, 100 pounds of giant powder, and 200 pounds of common powder, which had been stored in the bedroom of General J. L. Van Bokkelen. The General and nine others perished. It was supposed that the explosion was caused by a mischievous monkey that the General kept in his apartment. The General was an agent for explosives.

[13]*U. S. Mineral Resources* for 1872, p. 119 ; Yellow Jacket annual report, July 1, 1873. Burleigh drills were installed in the Burleigh Tunnel in Colorado in 1869, and were used in the East for quarrying in 1870. (*U. S. Mineral Resources* for 1871, pp. 487–492). Various early machines are described in *U. S. Mineral Resources* for 1869, pp. 503–517.

[14]The early diamond drills are described in *U. S. Mineral Resources* for 1869, pp. 518–526. The first diamond drill on the Pacific Coast was used on Telegraph Hill, San Francisco, early in 1870. (*U. S. Mineral Resources* for 1870, pp. 66, 67.)

lay ahead. When a heavy flow of boiling water spurted out, the hole was plugged and work in that direction abandoned. This occurred often in the deep levels.

LUMBER AND FIREWOOD—THE V-FLUME

The Sierras were devastated for a length of nearly 100 miles to provide the 600,000,000 feet of lumber that went into the Comstock mines, and the 2,000,000 cords of firewood consumed by mines and mills up to the year 1880.[15] In the early days, after the piñon pines had been cut on the Virginia and Como Ranges, the supply came from convenient timber on the lower slopes adjacent to Washoe and Carson Valleys. Gradually the lumbermen worked up to the crest of the range, then over on the west side. The magnificent forests surrounding Lake Tahoe constituted the major supply for years. No later visitor could conceive of the majesty and beauty fed into the maws of those voracious sawmills.

A large supply came also from the forests on either side of the Central Pacific Railroad after its construction, from what were known as Hobart's mills. The headwaters of the Carson River, 100 miles southward from the Central Pacific provided much of the firewood and some timber, which was floated down the river during the spring freshets.

The great invention of the V-flume for conveying lumber and firewood down the mountainsides was devised by J. W. Haines, who was lumbering in Kingsbury Cañon, back of Genoa, in 1866. It occurred to him to float the lumber down, and he made a box-flume for the purpose. The following spring he devised the very simple V-flume by nailing two planks together on their lower edges in V-shape, as men had been doing for centuries in smaller form to carry water. Haines patented the invention in 1871 and brought suit against Sharon and associates for infringement. The court held, however, that the patent was invalid because the device had been in common use for two years prior to the filing of the application.[16]

These flumes, planed on the inside, half-filled with water, and

---

[15]Lord's *Comstock Mining and Miners*, p. 351 (1883). He says this is "a careful estimate based on official reports of mining companies; of State Surveyor General; of U. S. Commissioner, and records of Virginia and Truckee Railroad."

Becker estimated the consumption of timber by the mines at 450,000,000 feet, and the firewood consumed by mines and mills at 1,800,000 cords. *Geology of the Comstock Lode*, p. 6 (1882).

[16]Lord's *Comstock Mining and Miners*, pp. 256–258 (1883).

on a fairly steep grade, carried a large quantity of lumber or firewood—as much as 500 cords of the latter in one day. When wood or lumber was thrown into the flume the water filled to the brim and the load floated free. A large number of such flumes, some of them many miles in length, were in use along the Sierras for years.[17]

The Bonanza Firm, in a quarrel with the Sharon interest over the price of lumber and firewood, bought a large tract of timber on the east slope of the mountains seven miles south of Steamboat Springs, built sawmills, constructed a V-flume 15 miles long, and supplied its own mines and mills. On the Comstock the lumber company was known as Mackay & Fair's. The Enterprise of March 31, 1875, reports that "Mackay & Fair's new wood flume at Huffakers on the Truckee Meadows will be completed about July 1." The net profits of that enterprise were only $645,030,[18] but the effect was to reduce substantially the prices of lumber and firewood.

The correspondent of the "New York Tribune,"[19] told of a visit to the lumber mills and surroundings with Flood and Fair, and of a fearsome thirty-minute ride down the fifteen miles of flume in what was called a "boat," which consisted of two twenty-inch boards nailed together in V-shape to fit into the flume, closed at the back and open in front, with strips of board 2½ feet long nailed across the top for seats. Part of the time the flume was near the ground, but much of it was on the top of high trestlework in order to keep the flume at a fairly even grade. Water sprayed on them from front and back. There was nothing to cling to but the seat and nothing but the blue sky above. "Flood said he would not make that trip again for all the silver and gold in the Consolidated Virginia."

---

[17]The lumbering industry on the Sierras is described and illustrated on pages 75–99 of John D. Galloway's *Memorandum* (1939). The Bancroft Library has a copy and the Mackay School of Mines another.

[18]Mackay and Fair Letters at Mackay School of Mines.

[19]*New York Tribune*, September 16, 1875.

CHAPTER XXV

Fire in the Stopes—Low-Grade Operations in the Bonanza Mines—
The Comstock Milling Monopoly—The Last Washoe Process
Mill—Losses in Tailings—Tailings Reworked.

FIRE IN THE STOPES

The immense quantity of timber used to fill the stopes of the
Con. Virginia and the California was often remarked upon:

> Every ton of ore extracted from the Con. Virginia and
> California mines leaves a corresponding vacuum. That
> space is filled with solid 14- and 16-inch timbers, leaving
> only a sufficient space between the huge bulkheads for
> the passage of men and cars. * * * The cost of these
> timbers at the mines is $21 per thousand feet (board
> feet), but even at these figures, it is much cheaper to fill
> with timber than to employ men to fill them with waste
> rock.[1]

Not less than 150,000,000 feet of timber, board measurement,
had been packed into those stopes and workings—enough to build
a dozen small cities—and a fire would turn the mines into a vol-
cano. Lord tells of that danger and of the vigilance of Mackay
and Fair.[2]

Fortunately, no fire occurred until May 3, 1881, when the
bonanza ore was exhausted. There was no hope of quenching it,
so all drifts and other openings into the stopes were closed and
sealed in order to shut off the supply of oxygen. Three years
later, when the fire was brought under control by the injection
of carbonic acid gas, the upper stopes were opened and the extraction of low-grade ore was begun. Meantime, the bonanza mines
had been levying assessments to carry on deep mining. That
hope failed at the end of 1884, the pumps were drawn, and the
lower workings began to fill with water. The shares of Con.
Virginia and California, which had already fallen to 15 cents,
then sold at 5 cents. The Bonanza Firm had not given the stock
any support for years and the speculative public lost all interest
until the low-grade operations proved unexpectedly profitable.

---

[1]*Engineering and Mining Journal*, Vol. 23, p. 262; *Mining and Scientific
Press*, March 31, 1877; S. P. Dewey's *Bonanza Mines*, p. 55 (1878).
[2]*Comstock Mining and Miners*, p. 321 (1883).

Low-Grade Operations in the Bonanza Mines

In 1883 Senator J. P. Jones, who had been mining low-grade ores from the old stopes of the Crown Point and the Belcher for three years (as a lessee) was given a lease on the Con. Virginia stopes from the 1550 level upward under an agreement to pay a royalty of 50 cents a ton for every ton milled. All of the openings into the stopes had been sealed since the fire broke out in 1881 and it was stipulated that he should not begin operations until the stopes could be entered.[3]

Mackay was in Europe practically all of that year engrossed in the affairs of the proposed Atlantic cable, and it is evident that neither he nor Superintendent Patton had much confidence that the fills and margins of the old stopes could be mined at a profit. All of their efforts during the preceding four years had been spent on a search for a new ore body below the Con. Virginia bonanza. Development work down to and including the 2900-foot level had been a continual disappointment, and on January 1, 1885, deep mining in the North End mines was abandoned. Ten months later the water was at the 2000-foot level and still rising.

Patton notified Jones in the spring of 1884 that he had extended a drift into the stopes on the 1200-foot level and that he could begin operations. Jones commenced in May, and up to November 1, 1885, had mined and milled 18,487 tons of ore yielding $310,109.69, or $16.70 a ton, valuing silver at $1.2929 an ounce. The discount brought the value down to $14 a ton.

As soon at it appeared that Jones was succeeding the Con. Virginia company began to extract low-grade ore below the 1550-foot level. For economy of management and operation the Con. Virginia and California companies were reincorporated on October 1, 1884, as the Consolidated California and Virginia Mining Company with a capital of 216,000 shares of the par value of $100 each. The company itself mined 19,670 tons, yielding $15.91 a ton during the first year, which gave a small profit. Mackay wanted the company to take over all of the operations and he

---

[3]When the fire burned out the millions of feet of timbers which had been packed into the stopes as the ore was removed, the whole country caved downward to fill the vacancy. The cave extended far up on the hillside back of the town leaving a long crack like an earthquake slip. So great was the pressure in the stopes that pieces of old 14-inch timbers were compressed to 6 and even 4 inches and resembled petrified wood. The town itself slid downward a little, but without damage except to brick buildings. *Nevada Historical Magazine* for 1911–1912.

persuaded Jones to surrender his lease to the company by agreeing to give him a one-third interest in the new milling company to be organized to mill the ores.[4]   James L. Flood, who had taken his father's place in connection with mining affairs, was the third partner.[5]

The Jones lease was surrendered on January 1, 1886, and the Consolidated Company entered upon ten years of very profitable mining in and about the old stopes, although the operation would have been far less successful except for the lucky discovery of three narrow sheets of good ore adjoining the old California stopes.   The first was found in the summer of 1886, the second in 1891, and the last in 1894.   It happened that the first was encountered after Mackay returned to take charge while Superintendent Patton took a vacation.   Fair had done little crosscutting on either side of the bonanza owing to the rush of water that followed the cutting of clay walls.   In these later years the stopes were practically dry as the water had been drained by deeper workings.

Superintendent Patton in his annual report of October 1, 1886, tells of the discovery of a body of good ore standing 65 feet east of the old California stopes on the 1400-foot level, which, he says, "gives promise of being important."   That ore was found unexpectedly in a crosscut that was being driven to facilitate the extraction of low-grade ore.   The discovery brought about the last disastrous stock boom, known as "The 1886 Deal," which was designed to punish the brokers and speculators who had been making a living by "shorting" stocks, but hurt many others.

The report for 1887 states that the new ore body stands parallel to the old California stopes, that it is 400 feet in length, from 5 to 50 feet in width, and extends from the 1250- to the 1500-foot levels.   The production for the year was 125,876 tons, yielding $2,969,555, nearly all of which came from the new ore body.   The average recovery was $23 a ton, which was 84 percent of the assay value.   The milling charge was $7 a ton.   Dividends for the year totaled $1,080,000, or $5 a share.   Mackay was struggling to save the Nevada Bank in the fall of 1887, and the

---

[4]James E. Walsh.

[5]James C. Flood died in 1889 after a long and distressing illness with Bright's disease.

It was said that Mackay and James L. Flood bought all of the stock in the treasury at the market price when these operations were begun.

increased value of his stock in the mine and the dividends were helpful. His ready cash was low because of the millions he had put into the Atlantic Cable and the Postal Telegraph. It did not become necessary to sacrifice any of his real estate in San Francisco and New York, nor his valuable railroad stocks.

Superintendent Patton left for Australia in October 1887 to become manager of the great Broken Hills mine, and was succeeded on October 1 by D. B. Lyman, who continued in charge of the mines and mills controlled by Mackay and Flood until after they withdrew from the Comstock in 1895.

Another large production was made during the year ending October 1, 1888, from which $1,080,000 was paid in dividends. Superintendent Lyman reports that a surprising quantity of good ore was found in the fills of the old stopes and in the adjacent walls while working over that part of the stopes originally mined in the years 1875 and 1876. This was due, he suggests, to the haste with which the rich ore was extracted. Both the yield and the dividends now decreased rapidly.

The last profitable year ended October 1, 1895, when $216,000 was paid in dividends from a small lens of ore discovered on the 1650 level in California ground. Mackay then surrendered control to some San Francisco stock brokers and retired from Comstock mining. His last visit was made in August 1895 in company with D. O. Mills.

During the years 1884 to 1895, inclusive, the mine produced 860,661 tons of ore, yielding $16,447,221, coin value, or $19.11 a ton, from which dividends amounting to $3,898,800 were paid, after the payment of $1 a ton royalty to the Sutro Tunnel Company. The value of the gold exceeded that of the coin value of the silver by nearly $2,000,000. The average milling charge was $6.50 per ton, with an 80 percent recovery. Mackay and Flood had large idle mills at that time, which enabled them to make a low milling charge. Mackay practically invited a test of the legality of that milling contract, but no stockholder brought suit.[6] The stopes on all of the levels from the 1200 to the 1750 were much enlarged during this period.

It is interesting to note that the low-grade operations in the bonanza mines yielded more in dividends than were paid by any of the other Comstock mines in all their history with the exception of three—the Savage, the Crown Point, and the Belcher.

---

[6]The company was well within the milling rules as laid down by Chief Justice Beatty in the Hale & Norcross case, in 108 *California Reports*, pp. 369–431, and had made a record for fair-dealing unmatched in Comstock history.

THE COMSTOCK MILLING MONOPOLY

The costly and unprofitable mills of the Ophir, the Gould & Curry, the Savage, and the Mexican during the early '60s caused those mines to send much of their lower-grade ore to custom mills. But those mills ran into debt during the lean years, chiefly to the Bank of California. When the time was ripe Sharon's Union Milling Company took them over and a new system was created whereby the productive mines ceased to own their own mills, except in small part, and had their ores reduced in mills belonging to the men in control of the mines. Thereafter, throughout the later history of the Comstock, the example set by Sharon was followed by Jones and by the Bonanza Firm, who controlled the producing mines which were not in Sharon's hands. The latter's milling rates were excessive as a rule, while those of Jones and the Bonanza Firm were moderate. Sharon and Jones, however, were not content to take their toll from profitable ore; when that failed they milled over 700,000 tons of low-grade ore during the '80s and '90s for the sole advantage of their mills. "Everything is arranged to suit the mills," wrote a correspondent. "The abuses are notorious, yet the local papers say nothing. Poor ore is mixed with good ore to increase the tonnage to the mills and there is little or no check on the sampling. Until the same respect is paid for mine stockholders as is now given to the mill stockholders your readers can expect no dividends from the Comstock mines."[7]

The system of milling ores in the private mills of insiders came to be regarded as a matter of course, if not of right, not only by the participants but by the general public, although after the "San Francisco Chronicle" and Dewey began their attacks there was widespread public criticism. Sharon and Jones took the precaution to have the names of others appear as trustees of their mines and their private milling companies, while Mackay and his associates acted openly and became members of the boards of their concerns.

The Bonanza Firm adopted the system when it took control of the Hale & Norcross in 1869. Mackay and Fair had two idle mills at the time, which they wanted to put in operation. Additional mills were acquired as more ore was developed, and, when that mine began to fail, the Firm took a gamble on the Con. Virginia, partly in the hope of finding some low-grade ore in the old upper workings for their idle mills. That hope was not realized, but

[7]*Mining and Scientific Press* of May 3, 1890, p. 305.

the lucky discovery of ore of moderate grade on the 1200 level soon put their mills at work. As the bonanza developed more mills were built or purchased, with the result that all of the bonanza ores were worked by the Firm, chiefly in large low-cost mills.

The charges for the high-grade ore were so moderate and the recovery so satisfacory that a court, under the rule adopted in the Hale & Norcross case, would have held the contracts reasonable. Nevertheless, the point remains that large profits would have been paid to the stockholders in dividends if the companies had owned their own mills, in which case the Firm, as the largest stockholder, would have received not less than one half of such dividends.[8]

We have an exact statement of those profits in a private memorandum made by J. Minor Taylor, the efficient office manager in Virginia City, to Messrs. Flood and Fair on September 11, 1881, which turned up in the Mackay and Fair files at the Mackay School of Mines. Taylor explains in detail the milling accounts from 1873 to the date of his report, which shows a net profit of $9,070,726.47. Fair was withdrawing from the Firm and they were having a settlement among themselves of the affairs of the Pacific Mine & Milling Company, which had carried on their milling business. Fair wrote back concerning some small items, which Taylor explained in a letter dated September 28. The expense for quicksilver is enormous, and the construction account is charged with $2,260,387.59.

As they had milled 809,275 tons from the Con. Virginia, 589,196 tons from the California, and 83,836 tons from the Ophir, Union, and Sierra Nevada—1,482,307 tons in all—the net profit, including milling and the recovery from tailings, was $6.12 a ton.[9]

Mackay gave close attention to the mills, which were efficiently managed by D. B. Lyman, one of the ablest millmen on the Comstock. Fair, who was not given to praise, but always to exaggeration, said Lyman was the only honest millman on the Lode.

---

[8]It may be said that the milling system was merely the exercise of a questionable official prerogative when compared with the iniquitous practice of the railroads in granting secret "rebates" and "drawbacks" to the Standard Oil and other favored shippers in the '60s and '70s, as set forth in Mark Sullivan's *Our Times, 1900–1925*, Vol. 2, pp. 284–292. The latter "system," he says, was "characteristic of the current philosophy," and "in this state of business ethics of the time lies the chief justification of Rockefeller and his associates in the South Improvement Company."

[9]The average milling charge paid by the Con. Virginia was $12 a ton, that of the California, which came into production later, averaged $11.

The values left in the tailings have been misrepresented and misunderstood, owing to the Comstock method of reporting mill returns. The ore as it left the mine was roughly sampled and weighed or estimated, which became the mine's valuation, with 10 percent or more of moisture included, while the assays made at the mill were of battery samples, dried. The average recovery on Con. Virginia ore, based on dry samples, was 73½ percent, and on California ore 74 percent, which was equivalent to above 83 percent if calculated on wet ore. This was a high recovery on ore of that grade.

A letter from Virginia City to the "San Francisco Bulletin" of June 11, 1878, comments on the statement of the "San Francisco Chronicle" that the report of the Con. Virginia shows that only 73⅗ percent of the values were recovered in milling, and that the tailings must be worth about $27 a ton, since the assay value of the ore was $100. The correspondent says that the percentage of recovery is based on a dry sample assayed at the mill, while the tonnage is calculated on the wet ore as it leaves the mine, containing an average of 14.2 percent moisture; that 6 percent of the values is lost mechanically by the failure of the particles of amalgam and slimes to settle; that one fifth of the tailings escape the reservoirs; and that the average value of the tailings is $10, of which not to exceed one half will be recovered.

Mackay testified in the Hale & Norcross case in 1892, in which he had no interest, that he had sold a large quantity of bonanza tailings at $5 a ton. He never could be convinced that a stockholder had a right to complain of a milling charge so long as it was reasonable and a proper recovery made. Despite the earlier Dewey suits and criticisms, he again milled the low-grade ores extracted from the bonanza stopes from 1885 to 1895, in association with James L. Flood and J. P. Jones. Mackay's idea that a director has the right to deal with his company if the contract was fair is now the law in California, by an Act passed in 1933.

Lord enters into an elaborate discussion of the Comstock milling system with especial reference to the Bonanza Firm, which concludes:

> If the managers had the lion's share of the profits, they had also the lion's share of the risk and labor. These facts should be borne in mind in any fair criticism or censure of their conduct as trustees.[10]

---

[10]*Comstock Mining and Miners*, pp. 113, 128, 246–249, 288, 330, 331 (1883).

THE LAST WASHOE PROCESS MILL—ELECTRIC POWER

The last Washoe process mill on the Comstock and the first to use electric power was built on the Chollar in 1887 by Capt. J. B. Overton to work low-grade Chollar ore as well as ore from the fine little ore body found on the upper levels of the Hale & Norcross in 1887. The mill, called the Nevada, had 60 stamps, each of 800 pounds, 30 amalgamating pans, 15 settling pans, and 10 agitators, after which the tailings passed over a long line of blanket sluices to catch the escaping amalgam and other values. Each pan held 3,000 pounds of pulp and was charged with 300 pounds of quicksilver and small amounts of salt and sulphate of copper, and sometimes soda or caustic potash.

The power plant of the mill was ingenious, and perhaps the first of its kind; part of it provided by an 11-foot Pelton water wheel at the surface, driven by water under a 460-foot head, brought down from the Water Company's flume. After the water passed the surface wheel it was dropped down the shaft 1,630 feet in two iron pipes to run six small Pelton wheels on the Sutro Tunnel level, which drove six Brush dynamos, each separately, the current being transmitted over copper wires to the dynamos in the mill.[11] Both installations operated successfully as long as the ore lasted. Electric power to operate mines and mills was brought from the power plant on the Truckee River, forty miles away, in 1900. The large mills built in recent years to work low-grade ores have used the cyanide or the flotation process, or both. The Washoe process has passed out of use, and ball mills for pulverizing ore have taken the place of stamps the world over.[12]

LOSSES IN TAILINGS AND QUICKSILVER

The Comstock mines produced a little over $300,000,000 from 1859 to 1880, excluding returns from tailings, and it has been said repeatedly that an additional $100,000,000, or 25 percent, escaped in the tailings and was irretrievably lost. But that estimate appears to be excessive.

The total loss up to 1880 is estimated at $70,000,000, and could not have exceeded $75,000,000. The total amount recovered from

[11] Dan DeQuille describes the mill on pp. 74–79 of his *History of the Comstock Lode* (1889).

[12] The two great mills of the '70s were built by the Bonanza Firm to work ores of the Con. Virginia and the California. The Con. Virginia mill had 60 stamps and milled 280 tons in 24 hours. The California mill with 80 stamps reduced 360 tons.

tailings saved up to that time is estimated at $23,000,000, including some that were reworked later. Hague, in 1870, estimated the average recovery from tailings as $5.50 a ton. Until the cyanide process was introduced the millmen did not expect to recover more than 50 to 60 percent of the values in the "tailings," or sands and the slimes. Some rich tailings were reworked twice, or even a third time after cyanide came into use.

The total loss in quicksilver is startling. On an average it exceeded one pound for each ton of ore milled up to the present time, or about 14,000,000 pounds. At an average of 60 cents a pound the monetary loss was $8,400,000. Dan DeQuille estimated the loss at 7,344,000 pounds up to 1876, by assuming too large an average in the earlier years.[13] When he was writing his Big Bonanza in 1875 he stated that the loss of quicksilver in milling the rich ores from the Con. Virginia bonanza "amounted to $60,000 and $70,000 per month"—a loss of over three pounds for each ton milled. The price at that time exceeded $1 a pound, the ore averaged $100 a ton, and they were charging each pan, holding 3,000 pounds of crushed ore, with 300 to 500 pounds of quicksilver. Necessarily, the loss in slimes and quicksilver was heavy. Practically all of this lost silver and gold and quicksilver ran down the cañons into the Carson River, which many have dreamed of as another Pactolus from which great fortunes were to be won. Such an attempt was made many years ago, which resulted in failure. If the material was light enough to be carried down to the river in small streams, the major portion would be carried on indefinitely, particularly by the spring floods.

TAILINGS REWORKED

The first effort to rework rich slimes and tailings at the Gould & Curry mill in 1864 was not successful, and it was not until 1866, after Louis and Henry Janin and Ira S. Parke had solved the problem that it became the practice to save and work them. Meantime about $20,000,000 had flowed away and was forever lost. In 1864 the newspapers report the Carson River choked with tailings from the Gold Cañon mills. It became the practice to catch some of the sulphides in the tailings by wide, shallow, blanket-lined flumes set at a gentle grade, after which they were sluiced into reservoirs for retreatment.

---

[13]Dan DeQuille's *Big Bonanza*, p. 145.

There is a very extraordinary example of blanket concentration in the ravines extending from the mills at
Virginia City and Gold Hill. The tailings from these
mills, estimated at not less than 600 tons a day, are
allowed to run into Gold Cañon and Six Mile Cañon
where they are passed over a great length of blankets
from five to six miles in length in each cañon.

Large tonnages of tailings were flumed and impounded in the
main and side gulches, and even on the flats near the Carson
River, for retreatment. In 1896 Prof. Robert D. Jackson of the
University of Nevada mining school successfully applied the new
cyanide process in treating Comstock tailings around Washoe
Lake. In 1901 Charles Butters erected a large cyanide plant in
Six Mile Cañon to treat a quarter million tons of impounded tailings. Later he added stamps and successfully cyanided both Comstock and Tonopah silver ores, resulting in the general adoption
thereafter of this method of treatment in the new mills erected.

When ore was reduced in custom mills, the slimes and tailings
became the property of the mills, not as a perquisite, but in reduction of the milling charge. Rossiter W. Raymond, the ablest mining man of his period, wrote of the milling charges in 1868:
"The low rates of custom mills are made possible by the margin
of profit between the 67 to 72 percent which the mills guaranteed, and the 75 to 80 percent which they actually extract." That
is, the mills saved an additional 5 to 8 percent from the tailings.

Raymond criticizes Dr. J. P. Kimball and J. Ross Browne for
assuming that the measure of the accuracy of the Washoe process
is the 65 percent which the mills guarantee to the mines. The
mines calculated the percentage on the weight of the wet ore,
while the mills made their return on dry ore. He says that "The
Washoe process is not a rude and imperfect one in principle, but
scientific and remarkably successful—better for the circumstances
than any European process."[14]

---

[14]*U. S. Mineral Resources* for 1868, p. 54; Id. 1869, pp. 697, 698.

CHAPTER XXVI

Bonanza Production and Dividends—Bonanza Firm Played Fair—
Mackay as the Comstock Knew Him.

That great ore body in the Con. Virginia and California produced from 1873 to 1882 a total of $105,168,859, from which $74,250,000 was paid in dividends, equivalent to 70 percent, from 1874 to 1881, inclusive. This is shown in detail in the tables on pages 260, 261.

A fire broke out in the mass of underground timbering in 1881, and the stopes were sealed in a partially successful attempt to smother it.

In 1884, the mine was partially reopened, and the extraction of low-grade ores in and about the old stopes was begun. Thereafter, from 1884 to 1895, inclusive, 860,661 tons of ore were mined and milled, yielding $16,447,221, an average of $19.11 per ton, from which additional dividends were paid, amounting to $3,898,800.

Total dividends paid from the bonanza ore bodies, 1874–1897, $78,148,800.

Total yield of the bonanza ore bodies, 1873–1897, $121,805,681.

Additional production from tailings estimated at $14,000,000.

Total production, 1873 to 1897, $135,805,681.

From 1895 to 1920, the only dividend paid was the sum of $64,800 in 1901.

BONANZA DIVIDENDS, 1874–1882

Paid by the Consolidated Virginia:

1874—$324,000 per month; $3 per share on 108,000 shares; May 1 to December 31, $2,592,000.

1875—$324,000 per month; $3 per share on 108,000 shares; January and February, $648,000.

1875—$1,080,000 per month; $10 per share on 108,000 shares; March 1 to December 31, $10,800,000.

1876—$1,080,000 per month; $10 per share on 108,000 shares; January to April 1, $3,240,000.

1876—$1,080,000 per month; $2 per share on 540,000 shares; April 1 to December 31, $9,720,000.

1877—$1,080,000 per month; $2 per share on 540,000 shares; May 1 to December 31, $8,640,000.

CONSOLIDATED VIRGINIA PRODUCTION STATEMENT, 1873–1882

Year	Tons	Gold	Silver (at $1.2929 oz.)	Total	Bullion discount	Total (Coin value)
1873	15,750	$314,288.68	$331,293.45	$645,582.17	$15,582.17	$630,000.00
1874	89,783	2,063,438.13	2,918,045.92	4,981,484.05	135,790.00	4,845,694.05
1875	169,094	7,035,206.54	9,682,188.22	16,717,394.76	640,715.00	16,076,679.76
1876	145,466	7,378,145.36	9,279,504.11	16,657,649.47	1,342,035.71	15,315,613.76
1877	143,200	6,270,518.68	7,463,500.39	13,734,019.07	975,416.05	12,758,603.02
1878	123,624	3,770,007.98	4,226,745.13	7,996,753.11	546,794.36	7,449,958.75
1879	60,227	1,198,319.68	1,283,039.15	2,481,358.83	219,482.40	2,261,876.43
1880	55,315	1,045,413.92	711,122.57	1,756,536.46	99,603.65	1,656,932.84
1881	6,816	91,889.71	52,253.72	144,143.43	13,745.16	130,398.27
1882	100	997.98	903.33	1,901.31	175.52	1,725.79
	809,275	$29,168,226.66	$35,948,596.03	$65,116,822.69	$3,989,340.02	$61,127,482.67

Year	Dividends	Assessments	Yield per ton, coin value	Milling cost	Cost per ton	Yield percent
1872		$277,150.12				
1873			$40.00	$13.77		
1874	$2,592,000		54.00	13.77		73.5
1875	11,448,000		95.00	13.00		73.25
1876	12,960,000		105.31	12.41	$18.76	72.3
1877	8,640,000		89.00	10.00	19.93	72.1
1878	5,400,000		60.26	10.47	17.73	74.2
1879	1,350,000		37.55	9.00	17.04	75.0
1880	540,000		29.95	9.00		81.0
1881		162,000.00	19.12	9.00		
1882		324,000.00	17.25	9.00		
	$42,930,000	$763,150.12	$75.53			73.0

PRODUCTION STATEMENT OF THE CALIFORNIA, 1875–1882

Year	Tons	Gold	Silver (at $1.2929 oz.)	Total	Bullion discount	Total coin value	Yield per ton, coin value
1876	127,540	$6,490,380.64	$6,910,460.76	$13,400,841.40	$895,521.01	$12,505,320.39	$98.05
1877	213,683	9,384,050.66	9,535,844.32	18,919,894.98	1,039,947.18	17,879,947.80	83.20
1878	138,785	5,552,585.48	5,396,493.45	10,949,078.93	564,372.89	10,384,706.04	74.82
1879	64,044	1,333,511.39	1,243,461.31	2,576,972.70	273,308.51	2,303,664.19	35.97
1880	38,359	547,484.52	342,558.91	890,043.43	43,950.68	846,092.75	22.05
1881	6,175	82,146.10	39,961.16	122,107.26	10,104.91	122,107.26	18.13
1882	610	5,111.22	5,025.06	10,136.28	493.47	9,642.81	15.80
	589,196	$23,395,270.01	$23,473,804.97	$46,869,074.98	$2,827,698.65	$44,041,376.36	$74.75

Year	Milling charge	Percent of recovery	Average daily wage	Cost per ton	Dividends	Assessments
1876	$12.65	$73\frac{3}{8}$	$4.05½	$20.00	$8,640,000	
1877	10.39	$73\frac{3}{10}$	4.17¾	19.09	14,040,000	
1878	9.15	72½	4.11	19.53½	7,020,000	
1879	9.00	74⅝	4.03	17.07	1,620,000	
1880	9.00	75		17.60		
1881	9.00	79				$162,000
1882	9.00	74				486,000*
1883						324,000
1884						324,000
		74%			$31,320,000	$1,296,000

The percentage of recovery is based on assays of dry ore at the mill. The tonnage is estimated on ore as it leaves the mine containing an average of 12% moisture.

5,123 tons of ore, valued at $453,060.46, or $88.43 a ton, were produced during the latter part of 1875, but as settlement was not made until January 18, 1876, that production was included in the report for 1876.

Included in "cost per ton," of the Con. Virginia and the California are bullion and other taxes, legal expense, discount on silver, supplies, and various other items.

1878—$1,080,000 per month; $2 per share on 540,000 shares; January to May, $4,320,000.

1878—$540,000 per month; $1 per share on 540,000 shares; May and June, $1,080,000.

1879—$270,000 per month; 50¢ per share on 540,000 shares; for five months, $1,350,000.

1880—$270,000 per month; 50¢ per share on 540,000 shares; for two months, $540,000.

The Consolidated Virginia paid $42,930,000 in dividends out of a gross production of $61,127,482.67 (coin value).

Dividends paid by the California:
1876—$1,080,000 per month; $2 per share on 540,000 shares; May 1 to December 31, $8,640,000.

1877—$1,080,000 per month; $2 per share on 540,000 shares; January 1 to December 31, $12,960,000.

1878—$1,080,000 per month; $2 per share on 540,000 shares; January 1 to July 1, $6,480,000.

1878—$540,000 per month; $1 per share on 540,000 shares; July, August, and December, $1,620,000.

1879—$540,000 per month; $1 per share on 540,000 shares; January, $540,000.

1879—$270,000 per month; 50¢ per share on 540,000 shares; May, June, August, and December, $1,080,000.

The California paid $31,320,000 in dividends out of a gross production of $44,041,376.36.

THE BONANZA FIRM PLAYED FAIR

The members of the Firm played a fairer game than any other group in control of Comstock mines. Mackay and Flood believed they had dealt honorably by their stockholders, and that in their market operations they had not taken undue advantages. Men must be judged by their times.

They never took over control of a property except for the purpose of searching for ore, and relinquished it when that hope failed.

They did not "put up jobs" by spreading false reports of strikes nor by mishandling a mine in order to depress the stock.

They never milled low-grade ore at a loss for the sole advantage of their mills. Ores were milled at reasonable prices and the best possible recovery made.

They paid dividends whenever it was possible, and their mines contributed $80,000,000 out of the total of $125,000,000 paid by all of the Comstock mines in all of their history.

They did more extensive deep development work than any other group and contributed the most in the way of assessments.

Mackay, Flood, and Fair devoted themselves to the management of the mines, with only Fair drawing a salary, and that a moderate one.

It is true that when the Con. Virginia bonanza began to fail they sold to the public a substantial part of their shares in the bonanza mines, retaining less than a majority, but very few directors in our mining and industrial enterprises fail to unload in a falling market.

The total profits of the Firm and of all the individual members from all Comstock operations (including milling profits and dividends from the Water Company, amounting to about $12,000,000) did not exceed $62,000,000, divided about as follows: O'Brien, $10,000,000; Fair, $15,000,000; Flood, $12,000,000, and Mackay about $25,000,000. The estate left by O'Brien, who died in 1878, that of Flood in 1889, and Fair's in 1894, tend to confirm this estimate. Mackay had three eighths of the Firm and remained longest on the Comstock; Flood and O'Brien had three eighths, and Fair two eighths. All died before their time: O'Brien at 52, Flood and Fair at 63, and Mackay, the only temperate member of the Firm, at 71, of a heart attack.

MACKAY AS THE COMSTOCK KNEW HIM

Mackay always remained a somewhat remote figure. He walked alone, kept his own counsel, rarely spoke unless spoken to. Easy of approach and courteous, but quick to dispatch the matter in hand. He took no part in the social life and avoided public gatherings, although ever ready to contribute generously to any good cause.

He had no familiars—not even among his few intimate friends. Not that he stood on his dignity—he was always his natural self— but, like General Grant, "a certain something in him seems to have held even the most familiar at a distance."[1] A few of his old-time friends called him "John"; some others called him "Mackay," but most men felt impelled to address him as "Mr. Mackay." Eliot Lord wrote of him: "One who knew him much

---

[1] Owen Wister's *Ulysses S. Grant*, p. 26 (1900)—a little classic now out of print.

longer and better than I told me he was more difficult to sound than any man he had even known; 'beyond a certain depth he remains inscrutable.' "[2]

His character was as clear as the lines of the Washington Monument. Men always knew where he stood. So truthful that even cynical old John Kelly admitted, "John Mackay never lies." A thing was either right or wrong — there was no shading between. A biographer of Sir Walter Scott might have been speaking of Mackay when he said: "He was a man essentially simple, who kept his deeper feelings to himself, knew no smallness or meanness, loved simple things and simple people, and was of quite heroic courage."

Loyalty was second nature. He gave complete confidence and expected faithfulness—at times to his sorrow. His judgment of men was good, but not unerring, and he was the last to believe that a friend was untrue. So modest that he shrank from praise: "Anything approaching flattery was an affliction," wrote a friend of later years.[3] Yet confident in judgment, bold in execution, and inflexible in purpose. Absorbed in business, yet, somehow, finding time to think what he could do for others. "Always thinking of somebody else," as Dick Dey said. Devoted to detail, and at the same time planning greatly. No enterprise was ever finished before another was begun. He did not worry and fret about his business. He planned carefully, gave thought to every detail, and accepted the result. In him the Biblical admonition was exemplified: "See'st thou a man diligent in his business? He shall stand before Kings."

Mackay was of a nervous temperament, yet with a deep repose of manner unless aroused, when he was formidable. Visitors found him attentive, but brief. His mind worked so rapidly that decision was almost instantaneous. Like some other men of imperfect education his memory was tenacious, even of details. Matters of a private or personal nature he carried in his head.

After the discovery of the Con. Virginia bonanza, Mackay became the outstanding man on the Comstock. His name was constantly on people's lips—almost invariably with words of praise. Everything about him was distinctive: his modesty, his reserve, his unfailing kindness to old friends, his innumerable

---

[2] *World's Work*, Vol. 13, p. 8160.

James E. Walsh, who was one of Mackay's friends and business associates for thirty years, was surprised to learn from this writer that he was a Mason.

[3] H. Alloway in *New York Tribune* of July 27, 1902.

benefactions, his uprightness, and the simplicity and decency of his life. At the same time he was human enough to have some small vices. He smoked and drank in moderation, and swore freely when aroused.

Men respected Mackay for his character and his brains more than for his money. A desire for fame he would have thought childish. The love and the respect of men sufficed—and that he had in great measure. Publicity, that much-abused word, he despised. No biography of him was ever written; every sketch of his life that has been attempted is incomplete and imperfect because he would not talk of himself.* He was repeatedly urged to become a candidate for a seat in the United States Senate, but flatly refused.

Eliot Lord throws light on Mackay as a miner:

> Fortune was kind to him, but he left no stone unturned to achieve success. * * * His insight appeared extraordinary when his ventures turned to gold, but he possessed no divining rod except close observation, thoughtful consideration, and a swift grasp of opportunities. His cool common sense was a rarely erring guide. Though apparently a desperate gambler, he took no ill-considered risks, for he never professed to see further into the Lode than could be seen by the aid of pick and drill.[4]

Mackay told much of himself in a few words when he replied to the question of the New York World, "Does Wealth Bring Happiness?"

> I am surprised that any one should think for a moment that happiness depends upon wealth. I was very happy during my early struggles with poverty. I enjoyed the toil, privation, and hardship I endured to win wealth. When a laborer in a New York shipyard, when swinging a pick and shovel as a miner, I was as happy as I can ever be.
>
> I had faith in and hope for the future, and when I began to realize that hope, by working hard, saving my money, and watching my opportunities, what a happiness I experienced—such a happiness as the possession of my subsequent fortune has failed to give me. I must

*See footnote, end of chapter.

[4]*Comstock Mining and Miners*, p. 302 (1883).

therefore answer your question by saying that I do not think wealth brings happiness.

John W. Mackay.[5]

In earlier years, at the height of the Con. Virginia bonanza, he said to Robert L. Fulton: "A man who has $200,000 and tries to make more does not know what he is doing."[6]

Mackay's fondness for boys was a tradition on the Comstock. Perhaps the absence of his own little sons in Paris caused him to think oftener of boys in general. It was not his way to stop and talk, but rather to think what he could do to make them happier. If he attended the theater he would buy tickets for the boys that stood about the door. When a circus came to town Mackay was always on hand to see that every boy got into the tent. He would arrive with his overcoat pockets filled with silver dollars, and, slipping around among the boys, would press one into the hand of each. They soon learned to anticipate him and would hang about the ticket wagon in droves.

It was Mackay's custom to walk down the hill to the hoisting works every morning when the night shift came off. One morning his attention was attracted to a barefooted boy crying with the cold, and he gave him a five-dollar gold piece with instructions to buy a pair of boots—boys always wore boots in those days. Lo, and behold! the next morning the road to the mine was lined with barefooted children. Mackay's generosity was imposed upon a thousand times—without ever hardening his heart.

He was the hero of every boy that was reared on the Comstock. Years later one of those boys wrote: "The sight of Mr. Mackay would set us small boys all aglow. We thought he was one of the wonders of the world."[7]

A dispatch from Virginia City, printed in the "San Francisco Bulletin" of July 21, 1902, says: "Among those who feel most keenly the death of John W. Mackay are the thousands of young men whose boyhood days were spent in Virginia. They always looked upon him as the kindest man in the world."

Every boy wanted to grow up "to be a man like Mr. Mackay." While none of us even approached the ideal, the ambition was a stimulating one throughout life to many a Comstock boy.

Mackay was one of the few wealthy men of the Comstock that

---

[5]*New York World*, July 27, 1902. The article does not give the date of the letter.

[6]*Nevada Historical Society*, Pub. 1909, p. 85.

[7]Jerome J. Quinlan, a fine character who lived all his life in Virginia City and served for many years as County Recorder.

riches did not corrupt or steal away his good name, but rather served as his means to further develop American resources and to brighten the lives of those less fortunate.

Mackay might have become one of the rich idle men of the world if he had so wished. His Comstock fortune came at a favorable time for growth through investments in real estate, and in railroad, telephone, and utility stocks; however, he was not content with riches alone.

He never enjoyed talking over money and investments as Americans are wont to do. He, as always, avoided the feverish excitement of gambling in stocks. He had enjoyed the battle to make his mining ventures successful and upon leaving the Comstock he plunged into the great business adventure of a new Atlantic cable, then to a new network of telegraph lines, the Postal, and death overtook him as he was about to span the Pacific with its first cable.

He demonstrated that his success on the Comstock was due not only to good fortune, but in great part to his own native ability, industry, and good judgment. He was able to hold his own with the sharp financiers of New York and London.

It was only natural that his son, Clarence H. Mackay, worshipped his father, and gave abundantly to the University of Nevada and its Mackay School of Mines to keep before the youth of Nevada, as an inspiration, the figure of John W. Mackay.

Upon the campus of the University is the bronze statue of John W. Mackay which stands in front of the Mackay School of Mines at Reno, Nevada. The head was modeled from a photograph taken when Mackay was forty-four years of age. The sculptor was Gutzon Borglum, one of America's most distinguished artists. (See frontispiece.)

Mackay's many visits to Europe began in 1871 when he first took Mrs. Mackay to Paris. Two years later he returned to bring her back. After the family removed to Paris in 1876 he made yearly trips, during which he traveled about Europe a great deal with chosen friends, visiting art galleries and listening to good music—for they were his principal enjoyments.

The one thing that fortune failed to bestow upon Mackay was an early education—to his lifelong regret. Nor did he ever come to realize how far he had overcome that by years of experience in the handling of great affairs, by constant contact with cultured men, by reading and writing and travel. The modesty of the man ever caused him to think lightly of himself. He was as familiar with Europe as America, was a lover of music and the drama, as he had always been on the Comstock, and became much interested in good pictures and works of art.   His tastes and thoughts were of a high order, and his naturalness and poise enabled him to meet with superior people on equal terms, although others were always encouraged to do the talking.   That Irish accent remained with him to the last.   "Not a common brogue," as Walsh said, "but the speech of a better class Irishman."   He had conquered poverty, overcame lack of education, mastered defective speech, and won the regard of the world.   He was proud of his wide acquaintance with important men of the day, and covered the walls of his office with their inscribed photographs,[8] but his heart remained in the West.   He used to say that manhood was at its best beyond the Rocky Mountains.

*A biography of Mrs. John W. Mackay by Ellin Berlin, the Mackays' granddaughter, contains much material on Mackay's life and character. Entitled *Silver Platter*, the book was published in 1957 by Doubleday & Co., Inc. (Editors, 1966.)

[8]Davis' *History of Nevada*, Vol. 2, p. 1066 (1913).

CHAPTER XXVII

Why Deep Mining Ceased—Cornish Pumps.

The last hope of finding a bonanza below the Sutro Tunnel level was abandoned when the Combination Shaft ceased pumping on October 16, 1886. All of the other deep shafts had already closed down one after the other. It had required all of the hopefulness of old-time Comstockers to carry on deep mining in the face of constantly increasing difficulties.

First—While the Lode had been developed on its dip eastward to the depth of nearly 4,000 feet[1] by incline and vertical shafts for a length of three miles—from the Sierra Nevada to the Overman—no bonanza had been found during the ten years of deep mining.

Second—The quantity of water and the temperature had constantly increased below the 1,000-foot level. The quantity was not the great problem, but the temperature. The forbidding question was whether the superheated water would not soon prohibit all mining at deeper levels.[2] Development work had been greatly restricted by the hot water encountered, and large areas were prospected only by diamond drills. Men had almost reached the limit of endurance.

Third—The difficulties and the expense of mining had increased in geometrical ratio with depth and nothing less than a large body of rich ore would be profitable.

Fourth—$40,000,000 had been expended in ten years by mines that did not pay a dollar in dividends. The prices of stocks had declined until the quotations meant little or nothing, and many stockholders were allowing their shares to be sold for the amount of an assessment. The report of the Director of the United States Mint for 1882 suggested: "It is a serious question whether it is advisable to continue operations in those mines. The ledge matter at these great depths (2,500 to 3,000 feet) is of great width

---

[1]Some of the third-line shafts intersected the Lode at the vertical depth of 3,000 feet, which was equivalent to over 4,000 feet on the dip. The Ophir, the Hale & Norcross, and the Crown Point-Belcher incline shafts followed down on the Lode to that depth.

[2]Becker said four years before deep mining ceased: "By far the greatest obstacle to mining on the Comstock has been the heat, which increases about 5 degrees Fahrenheit for every additional hundred feet sunk, and which seems likely eventually to put an end to further sinking." *Geology of the Comstock Lode*, p. 3 (1882).

but it contains only seams and stringers of ore, in most cases of very low grade."

Deep mining ceased first in the Gold Hill section in March 1882, when the New Yellow Jacket shaft was shut down after Mackay had failed to induce the adjoining and connecting mines to bear a share of the pumping expense. He was willing to continue notwithstanding the fact that there had been far less encouragement in the lower levels of the Yellow Jacket than in the other mines. He may have thought the mine had not been fully explored on the 3000-foot level and those above. To sink the shaft deeper, with water spouting from the drill holes at 170°, would have been next to impossible.[3]

The Belcher reports show that a considerable quantity of low-grade quartz was encountered between the 2200- and 2300-foot levels. Below that the reports do not speak of any encouraging discoveries except that a crosscut from the south drift on the 2760-foot level "passed through a mixture of quartz and porphyry which yielded some very good assays." A strong flow of water prevented further development. A winze below that level passed through porphyry and quartz for 75 feet, but was stopped by a heavy flow of water. Small bunches of ore were found occasionally on the deeper levels, and Becker says (p. 278): "Small stringers of good ore have been met on the 3000-foot level." Development work from the 2760- to the 3000-foot level was much restricted by heavy pressures of hot water, although drifts were run on both levels to connect with the New Yellow Jacket shaft.

The Crown Point in 1878 ran a drift to the Belcher pump shaft on the 2000-foot level in which "owing to intense heat and lack of ventilation men could work but a few minutes at a time." A 200-foot crosscut on the vein at that level assayed from $4 to $10 a ton—"very hot caving ground." From the 2000- to the 3000-foot levels the mine was developed by a winze from which connections were made with the Belcher pump shaft on the 2300-, 2500- and 2700-foot levels. Where heavy flows of water were encountered in diamond drill holes the drifts were bulkheaded.

---

[3]Mackay and his associates had taken control of the Yellow Jacket in 1876 in order to sink a great third-line shaft, in the hope of finding ore far below the old upper bonanzas. The shaft was one of eight vertical third-line shafts which were sunk at varying distances east of the Lode, and to varying depths. Mackay and his associates sank or joined in sinking five of them. No ore worth mentioning was found in any of them. The New Yellow Jacket shaft was managed by two rarely capable men: I. E. James and his successor Captain Thomas G. Taylor.

The Belcher and the Crown Point made connections with the New Yellow Jacket shaft on the 2300-, 2560-, 2760-, and 3000-foot levels, and the Yellow Jacket found itself compelled to pump much of their water owing to the frequent breaking of the Belcher pump rod. It was pumping also for the Exchequer, Imperial, and other mines. When the Yellow Jacket fly wheels were wrecked by a sudden fall of the great pump rod, on November 11, 1880, those mines filled to the 2828-foot levels in one day, and the Yellow Jacket was not pumped out again to the 3000-foot level until the following June. Meantime the Crown Point and the Belcher had pumped out their shafts to the 2760-foot level, but made no further effort. The water stood at that point in February 1882, when the Yellow Jacket pumps were again overwhelmed by a rush of hot water from the 2800-foot level of the Exchequer.[4] The Yellow Jacket kept its pumps running and notified its neighbors that they must contribute to the pumping in the future. Weeks of negotiations followed, which resulted in failure, whereupon the shaft was closed down on March 28, 1882, and deep mining ceased in the Gold Hill section. Evidently the managers of the Crown Point and the Belcher had surrendered all hope of finding ore below. They had already returned to their upper levels to mine the low-grade ore remaining in the margins of the bonanza, which was more to the advantage of their mills.

Captain Taylor returned to the old upper levels of the Yellow Jacket and mined low-grade ore at a profit during the following year, but Mackay and Flood were not interested in that and surrendered control of the mine to "Bob" Morrow, for the Sharon interests, which continued to mine large quantities of low-grade ore for years for the benefit of their mills.[5]

The North End Mines (the Con. Virginia, California, Ophir, Mexican, Union, and Sierra Nevada) were the next to stop pumping—at the end of 1884. The four last-named mines had been under the control of the Bonanza Firm for six or seven years, and the whole group was operated and connections made as if it were one extensive mine. After the Lode proved barren to the 2500-foot level in that group, all development work was done

---

[4]Superintendent Taylor reported that the flow was 100 miners' inches (1200 gallons a minute), and came so suddenly that he barely had time to get his men out of the 3000-foot level of the Yellow Jacket. The temperature of the water was 157 degrees Fahrenheit.

[5]When Mackay and Flood surrendered control of the Yellow Jacket to Morrow in July 1883 (Fair had withdrawn from the Firm in 1881), the mine had 120,870 on hand (the profits from milling low-grade ore), and had declared a dividend, which Morrow promptly rescinded.

in the shattered hanging-wall country along the northeast rift or shear zone. The plan of operation was not to sink the main shafts deeper, but to drive crosscuts eastward four or five hundred feet into the hanging wall and make new connecting levels between the mines, from which winzes were sunk to the depth of a few hundred feet, where new north and south levels were extended eastward. In this way all of that eastern hanging-wall country was gridironed with workings from the 2500- to the 3100-foot levels without finding a ton of payable ore. It was costly mining, as water and waste were handled again and again before reaching the main shafts. There were no such shear zones or rifts in the hanging wall from the Con. Virginia southward, and the development work in all of the mines from the Gould & Curry to the Overman and the Caledonia was confined to the main Lode.

The Con. Virginia and the California reports in the early '80s tell of cutting seams of quartz on the 2600- and 2700-foot levels "giving assays," without stating values. Supt. Patton says: "By the judicious use of diamond drills I have been able to avoid or control the influx of larger quantities of water than our pumping machinery can handle." The reports continue to speak of encountering and avoiding areas of hot water. When the joint Con. Virginia and California winze reached the 2900-foot level in 1884 a crosscut was extended 300 feet easterly and "diamond drill holes run ahead which showed indications of a strong flow of water, whereupon work was suspended." A crosscut was then extended 200 feet westerly and a south drift started, "but the water proved to be too strong." All of the water referred to was very hot, but no temperatures are given. That was the last work done on the lower levels of the Con. Virginia and the California.

The reports of operations in the Sierra Nevada, Union, Mexican, and Ophir after 1881 read much the same. Superintendent Patton writes encouragingly of occasional bunches of low-grade ore and "of narrow streaks of gold-bearing quartz giving high assays from selected samples, but averaging low in value." The following is not an unusual comment concerning operations below the 2500-foot level:

> By the use of the diamond drill at points where indications of water were found, I have been able to avoid large and heavy influxes which would seriously interfere with our working, and by controlling the same when found by a system of pipes, plugs, and valves, have

enabled our pumping machinery to handle it without difficulty.

There was a great width of mineralized material in that shattered northeast country but no ore. Drill holes on the 2900 and 3100 levels "showed much water." In June of 1884, after the four mines had been connected on the 3100-foot level, the Sierra Nevada and the Union ceased operations and bore a portion of the expense of sinking the Ophir-Mexican joint winze to the 3360-foot point. Some quartz was found in the winze, but when diamond drill holes were sent out in every direction from the bottom without encountering anything but strong flows of hot water, the last hope was surrendered and the North End mines ceased pumping at the end of 1884.

It must have been hard for Mackay and Flood to admit defeat. The former had said toward the end of 1880 that he believed those sections of the Lode that have proved fertile will yield bonanzas at some point below: "Although we may not get these bonanzas in the next five years, still they will be found."[6] Flood was equally confident three years later when, on a visit to the Comstock, he said that he and Mackay had surrendered control of the Yellow Jacket to Morrow after deep mining had failed there, but that they still held the North End mines. When the reporter called Flood's attention to Fair's statement to a New York reporter that the Comstock was finished, he replied: "Mr. Fair should not have said what he did. If he has not been misrepresented in what he did say, he did wrong. The fact is Mr. Fair is not a well man. He is ill. Six months in Europe may cure him."[7] Flood was a gentle man and not given to severe criticism.

The Alta shaft below Gold Hill and the Forman shaft on the hills to the eastward were shut down in December 1884. The Alta, which was planned to intersect the Comstock Lode at the depth of 4,000 feet, was only 2,150 feet deep. It had encountered considerable water and had found some ore on the 1950-foot level bearing heavy percentages of lead, which made a stir in the stock market for a time. The Forman shaft, locally known as "Forman's Folly," was a joint undertaking of the Overman, Caledonia, and Seg. Belcher. It was an ambitious effort, designed

---

[6] *Mining and Scientific Press*, October 9, 1880.

Mackay had said to Lord in 1880: "The Comstock is a poor man's pudding just now, but there may be more plums in it than we know of." *World's Work*, November 1906, Vol. 13, p. 8160.

[7] *Mining and Scientific Press*, June 23, 1883.

and equipped to intersect the Lode on its easterly dip at the vertical depth of 4,500 feet, but was stopped at 2,200 feet. The
superintendent was Charles Forman, an able man who came to
the Comstock in 1860 as joint agent for Wells Fargo Express[8]
with Captain Simmons.

Only the Middle Mines (Best & Belcher, Gould & Curry, Savage,
Hale & Norcross, Chollar, Potosi, Julia, and Bullion) then continued pumping. The Gould & Curry and Best & Belcher had
united in sinking the Osbiston shaft 3,000 feet to the eastward.
The Bullion and the Julia joined in sinking the Ward shaft,
which stood by the side of the railroad east of the Bullion. The
others had united in 1875 in sinking the great Combination shaft
on the hillside 3,000 feet east of the Chollar.

One marvels at the optimism and daring of those old-timers in
undertaking that row of third-line shafts from the Union to the
Alta, even while recalling the fact that the general public was
contributing the larger part of the expense.

The Middle Mines and those in Gold Hill had continued to sink
their second-line incline shafts and to explore the Lode on level
after level as far as permitted by the difficulties with water and
heat. The third-line shafts were designed to relieve them of
those conditions and to insure the deeper development of the
Lode. The heavy assessments levied during that period by each
of the leading mines—ranging from $250,000 to $500,000 a year—
were due to the continuance of operations in both second- and
third-line shafts.

The Savage ground was so wet that its shaft was not sunk
below the 2800-foot level and it was never prospected below.
Three separate crosscuts from the Combination shaft on the
2500-, 3000-, and 3100-foot levels had to be stopped and bulkheaded so great was the flow of hot water. On the 2500-foot
level some seams of good ore were found.

The Hale & Norcross, after making connections with the Combination shaft, was enabled to sink its incline shaft to the depth
of 3,200 feet. Small bunches and streaks of ore were reported
on the 2900-foot level and on the levels below, which were greatly
magnified in the newspapers. In the summer of 1885 there were
numerous reports of the discovery of a substantial body of good
ore on the 3000-foot level. The stock doubled in price, but fell
back when the ore was found to be bunchy and of no practical
value. Other bunches and streaks were reported down to the

---

[8]Thompson and West *History of Nevada*, p. 582 (1881).

3200-foot level where the last connection was made with the Combination shaft.

No ore was found in the Chollar and the Potosi from the 500- to the 3100-foot levels. The Bullion and the Julia never had a ton of ore; their joint Ward shaft reached the 2525-foot level where it encountered such pressures of hot water and so much caving and running ground that no attempt was made to sink deeper.

The Combination shaft intersected the Comstock Lode at the depth of 3,000 feet and entered a body of low-grade quartz on the 3200-foot level, which proved of no value. The shaft was then sunk to the 3250-foot point. The double line of Cornish pumps was unable to handle the water when the shaft began to make connections with adjoining mines, and Superintendent Requa installed a hydraulic pump to assist, using water furnished by the Water Company as a plunger. Later, two additional hydraulic pumps were installed.[9] The pumps were then lifting 5,200,000 gallons every 24 hours to the Sutro Tunnel level,[10] or 3,600 gallons a minute. This quantity lifted 3,200 feet would require about 3,000 horsepower theoretically, or with pipe friction and modern pumps and motors about a 4,000 horsepower continuous load.

On October 16, 1886, the Combination pumps ceased to operate. "Within 36 hours after the hydraulic pumps were stopped the water had risen to the 2400-foot level, filling the entire lower workings of the Chollar, Potosi, Hale & Norcross, and Savage mines, including several miles of crosscuts."[11]

The vertical-longitudinal section of the Comstock Lode covers all of the important productive mines. It does not include the Overman and the Caledonia, lying south of the Belcher, which produced considerable ore of fair grade from scattered ore bodies; nor the mines lying north of the Ophir, namely, the Mexican (new), the Union, the Sierra Nevada, and the Utah. The Mexican found a small ore body in 1911, at the depth of 2,400 feet; the Union and the Sierra Nevada produced about $5,000,000 from two adjoining ore bodies between the 2200- and 2600-foot levels, from which no dividends were paid, and the Utah was barren to the deepest levels. (See illustration, p. 276.)

---

[9]*Mining and Scientific Press* of October 23, 1886, says that the hydraulic pumps on the 2400-foot level drained the shaft to the 3200-foot level, and that the Cornish pump on the surface lifted the water from the 2400 level to the Sutro Tunnel at 1,600 feet.

[10]Dan DeQuille's *History of the Comstock Mines*, p. 88 (1889).

[11]*Mining and Scientific Press*, October 23, 1886.

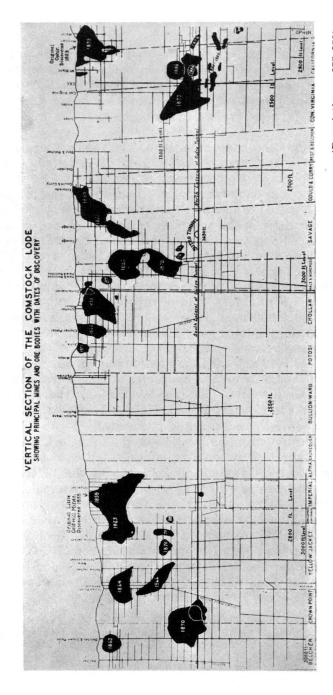

VERTICAL SECTION OF THE COMSTOCK LODE
SHOWING PRINCIPAL MINES AND ORE BODIES WITH DATES OF DISCOVERY

(Description, p. 275–278)

The ore bodies shown in outline on the section often represent several parallel or coterminous deposits. That is especially true of the Original Gold Hill bonanza, and of the ore bodies shown in the Chollar-Potosi, the Hale & Norcross, the Savage, the Gould & Curry, and the Ophir. The Chollar-Potosi found ten successive parallel or coterminous ore bodies scattered through a wide lode over a length of 1,200 feet and a width of 300. Unfortunately none of them extended below the 500-foot level.

The Con. Virginia-California bonanza is not large in outline but averaged about 77 feet in thickness. The two later parallel sheets of ore were found in 1886 and 1894 during the low-grade period. The small ore bodies trailing out below the bonanza were found in the northeast rift from 1900 to 1911 when the North End mines were pumped out to the 2500-foot level.

It will be noted that the ore bodies occur in three large groups, indicating that each was mineralized from the same deep-seated source.

No profitable ore was found in the Lode except in connection with the ore bodies outlined on the vertical section, although the extensive bodies of quartz and vein porphyry, both on the upper and lower levels, were never quite barren of silver and gold.

The section shown is a copy of the Atlas sheets accompanying Becker's "Geology of The Comstock Lode," published in 1882, and with some later additions.

Only a small part of the mine workings could be shown on this section, which exhibits the main shafts and the principal north and south drifts.

The Ophir was the most northerly bonanza and stood alone except for a narrow parallel sheet of rich ore lying about 40 feet to the west. South of the Ophir the Lode was barren for a length of 2,000 feet to the Gould & Curry bonanza, and remained so to the deepest levels. While the Con. Virginia bonanza was found below that section many years later, it did not lie in the Lode itself but in a vertical rent in the hanging wall 1,200 feet below the surface. Southward from the Gould & Curry to the Chollar-Potosi, inclusive, the parallel or coterminous near-surface ore bodies were almost continuous for a length of 3,000 feet, but, strange to say, no profitable ore was ever found below them. Then came a barren stretch of 1,800 feet of the Lode, including the south end of the Chollar-Potosi, the Bullion, the Exchequer, and most of the Alpha, which remained barren to the deepest levels.

The Original Gold Hill bonanza, including a small portion of the

Alpha, consisted of the Old Red Ledge, which dipped west, and the east ore bodies lying along or near the hanging wall. The east ore bodies formed a more or less continuous series for 2,500 feet along the hanging-wall side of the Lode from the Gold Hill bonanza to the Crown Point, inclusive. There the important upper ore bodies ended with the exception of a rich near-surface ore body of moderate size in the Belcher, which stood in the west vein.

The west vein was peculiar to the Gold Hill section; it was rich in the Little Gold Hill mines, barren through the Yellow Jacket until it reached the south end, where it made ore for 500 feet through the Kentuck and into the Crown Point, then barren for 700 feet when it made the upper Belcher bonanza. Beyond that, in the Overman, some small disconnected bodies of fair ore were found down to the 400 level.

It came to the surface, or nearly so, in the Little Gold Hill mines, where the vein and the ore terminated at 275 feet on a nearly flat bed of clay. It ceased in the same way in the Crown Point, Kentuck, and Yellow Jacket at the depth of 400 to 450 feet, but terminated upward in a thin blade of ore 90 feet below the surface.

It is remarkable that no profitable ore was ever found below the 1000-foot level in that rich section of the Lode from the Gould & Curry to the Overman with the exception of the great Crown Point-Belcher bonanza, which was discovered at the depth of 1,100 feet after the upper ore bodies were nearly exhausted.

CORNISH PUMPS

All of the Comstock mines employed the old Cornish pumps for handling their great volumes of water until Superintendent Requa installed hydraulic pumps to assist at the Combination shaft in the early '80s. While the pump reached its highest development on the Comstock it remained a huge cumbersome affair, costly to install although not expensive to operate except for the cost of fuel.

The last Cornish pump on the Comstock was installed by W. H. Patton at the Union shaft[12] in 1879. It had but one flywheel, 40 feet in diameter and weighing 110 tons, which helped to operate the pumping beam. The pumping engine was of the compound

---

[12]The Union shaft was a joint enterprise of the Union, Mexican, and Sierra Nevada mines.

Lord writes briefly on the great pumping and hoisting machinery on pp. 343–347, 367 of *Comstock Mining and Miners* (1883).

condensing, direct-acting type, with a capacity of 10 strokes a minute. The initial cylinder had a diameter of 64 inches, with a piston stroke of 6 feet 9 inches, and weighed 30 tons. The expansion cylinder had a diameter of 100 inches, a stroke of 8 feet, 3 inches, and weighed 43 tons. The pumping beam, 48 feet between centers, was built of sections of wrought iron with a double truss running around it, and was supported in the center on a steel shaft 22 inches in diameter. It was connected with the pump rod at one end, and operated by two expansion cylinders, one at each end. The pump rod was 16 inches square and 2,500 feet long, made up of lengths of Oregon pine strapped together with iron plates. As the pump rod plunged downward and forced up the water it lifted the pumping beam and lifted also the balance bobs set at intervals of 400 feet down the shaft, each of which required a long chamber in which to operate. Each bob consisted of a long beam balanced at the center like a see-saw, one end being connected with the pump rod and the other weighted with a heavy load of scrap iron. As the pump rod, weighing 300 tons, was lifted for another stroke, its weight was counterbalanced by the descending weights at the ends of the balance bobs, and by one of the expansion cylinders on the surface.[13]

All of the third-line shafts had four compartments (except the Forman which had five), one for the pumps, another for sinking, and two for hoisting. Every shaft had a sinking pump of the suction type in addition to a Cornish force pump. To overcome the great pressure, the water was not pumped directly from the bottom to the top of the shaft, but in stages, from one tank to another, the tanks being placed at the side of the shaft at intervals of about 200 feet. Each tank had a separate pump connected with and operated by the main pump rod.

When the huge iron plunger at the lower end of the rod descends into its barrel or cylinder it forces the water up the pump column, which is poured into the tank of the next pump above, and as the rod descends it draws the water from the tanks in which it stands. Thus a continuous stream is poured out at the top of the shaft. * * * The installation is calculated to operate a double

---

[13]In the earlier installations the pumping beam was not assisted by expansion cylinders, one at each end, but depended more upon gravity—a heavy load of scrap iron in a large box (called the pitman) was attached at the far end of the pumping beam, the weight of which helped to raise the pump rod after a plunge, assisted by similar balance bobs down the shaft.

line of 14-inch plunger pumps with a 10-foot stroke at the depth of 4,000 feet, and has a capacity of 2,000,000 gallons in 24 hours.

After the north lateral of the Sutro Tunnel reached the Union shaft the water was pumped to the tunnel level. The twelve boilers at the Union had a grate surface of 270 square feet, and consumed 33 cords of wood every 24 hours.

The 1,200-horsepower hoisting engine at the Union shaft was a double-cylinder, direct-acting, noncondensing engine, and, with the exception of a similar engine at the New Yellow Jacket shaft, the fastest on the Lode. Each cylinder had a diameter of 28 inches and a stroke of 8 feet.

The hoisting rope at the Union was of the flat wire-woven type, 4 by ⅝ inches, which had been generally used on the Comstock for many years. The machinery and equipment of one of those great hoisting works weighed over 3,000,000 pounds and, including foundations and the surface plant, cost not less than $1,000,000.

The hoisting and pumping machinery at the Union was designed to operate to the depth of 4,000 feet vertically, but the shaft was not sunk below the 2700-foot level. Below that point development work was continued to the eastward, in that shattered hanging-wall country, by means of crosscuts and winzes to the 3100-foot level.[14]

The pump at the New Yellow Jacket vertical shaft, 3,080 feet deep, had a capacity of 1,000 gallons a minute, or 1,440,000 gallons in twenty-four hours, and regularly raised over 1,000,000 gallons. The pump rod was 3,055 feet long, made of lengths of Oregon pine, 16 by 16 inches, strapped together with iron plates. Its weight when in motion was 1,510,400 pounds. Its greatest capacity was seven strokes a minute, each stroke lifting 160 gallons. The weight of the pump rod was equalized by 8 balance bobs placed at intervals in the shaft, carrying a total lifting weight of 240 tons. There were 13 pumps in the shaft, placed at intervals of about 250 feet, which lifted the water from station to station, all attached to the pump rod. The two flywheels weighed 125 tons.

The average monthly expense was $13,000, or 9 cents for each

---

[14]A main drift on the 2900-foot level was extended from the Sierra Nevada to the Con. Virginia, and the drift on the 3100 level was connected with the Ophir-Mexican winze.

ton of water lifted. The largest item was for fuel—24 cords of wood at $9.50 a cord were fed to the boilers ever 24 hours.[15] The water was pumped to the surface until connection was made with the Sutro Tunnel lateral on the 1500-foot level in March 1881.[16] Captain Taylor says that the pump rods in the Belcher and the Imperial incline shafts broke frequently, throwing the burden of pumping upon the Yellow Jacket, which was connected with all of the Gold Hill mines on the lower levels. He tells that when the Yellow Jacket pump rod broke suddenly on November 11, 1880: "The engine made a very quick up-stroke which broke both fly wheels and the upper and lower straps of the pitman, making a perfect wreck." Fragments of the cast iron wheels were thrown in all directions—some through the roof of the building. The water rose in one day from the 3,000- to the 2828-foot levels and pumping was not resumed for six months.

Rod catchers were installed at varying depths in all of the pump shafts, to prevent the rod from falling in case it broke. They were merely huge blocks of wood securely bolted and banded to the rod and would catch on the shaft timbers and prevent any section of the rod from falling more than a short distance. The pump shafts in all of the mines were larger than the adjoining hoisting shafts and the timbering more massive. The last Cornish pump to operate on the Lode was at the Combination shaft.

The pump rod was held in place down the shaft and vibration prevented by "stays," consisting of large sticks of timber placed on all four sides of the rod about every 30 feet, forming a square hole through which it played up and down. The stays and the rod were well greased and there was little friction. That great stick of timber was lifted and lowered 10 feet seven times each minute. At the same time the rod lifted the water in the 13 tanks up the shaft. This was made possible by the power of the flywheels and the counterbalancing of the pumping beam on the surface and that of the balance bobs down the shaft—and that cumbrous piece of machinery had to move like a watch.

It was thought remarkable that a three-decker loaded cage in the C. & C. shaft, running in guides, could be hoisted 2,700 feet in one minute.

---

[15]The Yellow Jacket had 10 boilers, each 16 feet in length, and 54 inches in diameter, with a grate surface of 250 square feet, which supplied steam to hoisting and pumping engines of 2,941 horsepower.

[16]Facts from Superintendent Thomas G. Taylor's diary.

A section showing the comparatively small pumping and hoisting machinery of a Comstock mine in 1870 is inserted opposite page 146 of Hague's able volume on the Comstock mines, Vol. 3 of *U. S. Survey of the 40th Parallel* (1870).

CHAPTER XXVIII

"The 1886 Deal"—The Revival from 1886 to 1894—The North End Mines Pumped Out 1899 to 1920.

This was rightly called a "deal" since it was a "squeeze" manipulated by James L. Flood (only son of James C.) and the managers of some of the other mines for the express purpose of punishing some of the few remaining brokers and small speculators who had been living for several years by "shorting" Comstock stocks.

When the reorganized Con. Virginia began to mine low-grade ore in and about the old stopes the price of the stock rose gradually to $2 a share, and then to $2.50 on October 7, 1886, after the annual report told of the discovery of a new body of ore east of the old California stopes. The superintendent stated that the find "gives promise of being important," but the speculators evidently discounted the news for the stock hung around $2.50 a share until the end of the month.

Some of the brokers had not only been "shorting" the stocks, by which they collected the regular assessments instead of letting them go to the companies, but others had been making it a practice to "bucket" their customers' orders.[1] To "bucket" an order means that the broker takes an order to buy a certain amount of stock for a customer on margin account and fails to purchase it. The customer deposits his margin in cash for about one half of the purchase price and gives his note to the broker, at not less than 2 percent a month, for the balance; the broker holding the stock as security. Now, Mr. Broker, feeling assured that he can purchase the stock if demanded at the same or a lower price, does not buy the stock but regularly collects the interest on the note from the customer and calls upon him for the assessments as fast as they are levied, although he does not pay them. Such transactions often ran for years with no demand from the customer to produce the stock. The practice not only depressed the market, but the companies failed to collect the assessments on perhaps a considerable number of shares which customers thought they had bought. King says that 81 stocks had been stricken from the board in 1881 for failing to pay their

---

[1] So this writer was informed by A. F. Coffin, who was a broker at that time and continued until his death three years ago.

dues, and that the remaining Comstocks were "selling for almost nothing" when the deal started.[2]

Flood's fellow conspirators, who controlled the Combination shaft, gave out word that the shaft was about to be shut down, which meant the end of deep mining, as the Combination was the last one pumping. Many that held stocks began to sell, others increased their short holdings. The shaft was shut down on October 16, but, instead of falling, stocks continued to rise. Flood and the others had been quietly buying, and continued to purchase for future delivery until they held nearly all of the stocks of several of the leading mines.

Then the market rose spectacularly. Cons. California and Virginia, the leader, jumped from $4.50 on October 31 to $12 on November 12, to $48 on December 2, to $55 two days later, and to $62 in one session. As usual, all other Comstocks rose in sympathy. Dey told H. L. Slosson that James L. Flood gave the shorts the opportunity of settling up several times while stocks were rising, but they declined on the theory that the market was about to break.

The poor deluded public, seeing the constant rise of the market, rushed in and began to buy. The "Mining and Scientific Press" of December 11 reported "that some $4,000,000 has been withdrawn from the savings banks in the last ten days, most of which, presumably, has been invested in mining stocks." The manipulators who controlled moribund and worthless mines went into action and a number of stocks that had been selling for less than $1 a share rose higher, proportionately, than Cons. California and Virginia.

People began to call upon their brokers for stocks which the latter, presumably, had bought for them, but the brokers could not deliver. The excitement was intense, threats were made, brokers' offices were in a turmoil.[3] Panic reigned on the Stock Exchange. A prominent broker in Virginia City failed, then

---

[2]King tells of that deal on pp. 300–303 of his *History of the San Francisco Stock Exchange* (1910).

[3]The Negro porter at the Nevada Bank had bought 1,000 shares of Cons. California and Virginia at low prices, and when the stock was high Dick Dey advised him to sell. When he gave the order his broker invited him into a back room for a confidential chat.

"Who told you to sell?"

"Mr. Dey."

"Why you poor sucker, don't you know that Dey advised you to sell so Mackay could pick up your stock?"

The stock was not sold and the porter lost every dollar.

several of the larger brokers in San Francisco, followed by a number of the smaller fry. They had been unable to fill their "shorts" and the customers whose stock they were holding, or supposed to be holding, were wiped out. King says that "Prices advanced so high one day and there was such difficulty in getting stock from the sellers that Mr. E. P. Peckham, President of the Board, arbitrarily closed the board for the day." Apparently the elder Flood did not approve of the slaughter: "Mr. James C. Flood returned from the East that night and we felt his hand the next morning. Consolidated Virginia broke to $32 a share and everything else tumbled with it."[4]

Nothing but death ever cured those old-time speculators in Comstocks.

THE REVIVAL FROM 1886 TO 1894

When the Combination shaft stopped pumping in October 1886, men thought the Comstock was finished. All that remained was the low-grade ore in the old upper levels which had been so honey-combed with workings that there was no hope of finding another bonanza, and only a chance of encountering some fair ore that had been missed. Nor did the remaining low-grade ore give any promise. Those old upper ore bodies had been stripped time and again of all rock that would pay a profit. The stock market was on its last legs.

The superintendent of the Cons. California and Virginia had announced the discovery of a new ore body on the 1400-foot level, but that was discounted. Then, like a bolt out of the blue, came the "1886 Deal," which created havoc among the brokers and the general public. But it soon developed that the low-grade operations in the Cons. California and Virginia were profitable and made especially so by the discovery of a parallel sheet of ore while running a drift to an old stope. Hope revived. The few remaining brokers and stock manipulators, who had thought their occupations gone were stimulated to carry on the same old game. Mines that had been closed down for years resumed operations on the old upper levels. The stock market was manipulated up and down, within narrow limits, and the lambs came back into the market to be fleeced in the same old way.

The Comstock revived and carried on hopefully for seven or eight years—although but a shell of the palmy '70s—due almost entirely to the encouragement given by the Cons. California and Virginia, which continued to mine a large tonnage of low-grade

---

[4]King's *History of San Francisco Stock Exchange*, p. 302 (1910).

ore and pay regular dividends. The Gold Hill mines had returned to the old upper levels after pumping ceased there, and were mining and milling large quantities of low-grade ore.

The Belcher and the Crown Point, controlled by the Jones interests, and the Yellow Jacket, by the Sharon interests, reduced 750,000 tons of ore averaging $12 a ton, mill returns, during the eight years following 1882, but only to the advantage of their mills. No dividends were paid, but on the contrary a few assessments were levied when the mill returns failed to pay the expense of mining and milling.[5] Such of the other mines as could find a little ore were producing on the same basis.

The Confidence discovered a little ore body on the 1100-foot level in 1888 which created a manipulated stock flurry. The yield was $900,000, from which only $124,800 was paid in dividends. A cave in an old drift disclosed this ore body. The Hale & Norcross stumbled upon an excellent little ore body in 1887 which extended from the 400- to the 700-foot levels. That was a perfect lens of $40 ore. This writer worked in the mine at the time. It produced $2,500,000, but those in control allowed only $168,000 to be paid in dividends.[6]

After 1890 the Comstock rapidly declined. The Cons. California and Virginia was paying smaller dividends, the Gold Hill mines had almost exhausted all of the ore that would pay the expense of mining and milling, and again the general public had grown tired of the game. Some mining and milling was carried on for several years by a number of the leading mines until 1895. Then followed several years of stagnation.

After 1881 some thirty mines were kept alive for the benefit of mill rings, politicians (local and national), salaries, and the remaining small stock brokers and manipulators in San Francisco—each mine with a superintendent and a full quota of officers. When the market was dull and assessments could not

---

[5]"Comstock mines during the past three years have been steadily increasing the yield from low-grade ores extracted from old workings in the upper levels; no dividends being paid, but nearly all steadily levying assessments. Of course there is no profit in the business on that basis," says the *Mining and Scientific Press* of December 27, 1884, "and yet most of those interested in the operations of these mines manage to get a profit out of them by ownership of the mills that crush the ore."

[6]The stockholders brought suit against the management for fraud in milling and recovered judgment.

It was said at the time that the ore was found by a miner while digging on the side of a drift to put in a set of timbers. Jimmy Fair scoffed at the first news of the discovery of a new ore body on the 700-foot level of the Hale & Norcross: "Why I swept the walls of that mine with a broom nearly twenty years ago!"

be collected the salaries would accumulate until a favorable turn.

There was little real mining done and hardly a pretense of economy in management. The small proportion of money paid for mining as compared with that spent on management is reminiscent of the "one half-penny worth of bread" which Falstaff provided for his journey as against "an intolerable deal of sack." The newspapers told of such conditions in a number of mines which applied as well to nearly all of the others that continued to operate.[7]

THE NORTH END MINES PUMPED OUT, 1899 TO 1920

In 1898 the Comstock appeared to be on the verge of collapse. The total production for the year was only $205,000, and that from ore which did not pay to mine and mill. Stocks had almost no value and the companies found it difficult to collect enough from assessments to pay salaries. Nevertheless, twenty-nine mines along the Lode were pretending to operate, each with a superintendent and a quota of officers and with more or less elaborate offices in San Francisco.

The brokers in control of the Virginia City mines were almost in despair when they thought up the scheme of reviving the stock market by pumping out the North End mines (the Con. Virginia, Ophir, Mexican, Union, and Sierra Nevada), and renewing the search for a bonanza on the deep levels. The plan met with enthusiastic approval of all the brokers and other interested parties. Even the bare possibility of finding a new bonanza would bring the speculators back into the market. The campaign was well handled. An advisory committee interviewed the mine superintendents, most of whom owed their positions to the brokers, to get their opinions of the project. They approved, of course, although a few conservatives insisted upon inserting qualifications in the signed report, which stated that owing to the difficulties of deep mining and the great width of the Lode there was not a deep level which "has been so thoroughly prospected that there is not yet a chance of finding ore in it." They recommended the effort provided the mines could be pumped "to the lowest levels" with a modern pumping plant "for one twelfth of the cost of pumping by the old system," as the promoters stated. That condition was impossible of fulfillment and the superintendents must have known it. When operations got well under

---

[7]*San Francisco Argus*, April 12, 1883; *Engineering and Mining Journal*, May 13, 1883; Whitman Symmes on *The Decline and Revival of Comstock Mining Mining and Scientific Press*, Vol. 97, pp. 496–500; 570–576.

way it was found that the cost of pumping at the C. & C. and Ward shafts with hydraulic and electrically driven pumps was as great as it had been to run two of the old Cornish pumping plants, although more water was raised for the short lift to the Sutro Tunnel. The water in the sump of the C. & C. shaft ranged from 130 to 160 degrees, and the difficulties with the pumps were endless.

The Comstock Pumping Association, which was to be supported by all of the mines, went into operation at the C. & C. shaft in 1899, after much alluring propaganda had been fed to the public and its interest excited by a manipulated stock market. The beginning was simple as the shaft was in good order to the water level, which stood 100 feet below the Sutro Tunnel.

There was little probability of finding profitable ore on the lower levels of the Con. Virginia, which Superintendent Lyman thought he had exhausted, and if none had been found within the next year or two when the difficulties and the expense had multiplied, the plan would have been abandoned. But luck was with the brokers. Soon after the water had been lowered below the 1800-foot level with a hydraulic pump a drift was extended northward along the northeast rift and the extraction of low-grade ore was begun. Then, almost unexpectedly, a rich little body of ore was encountered, and then another. Thereafter, from 1901 to 1913, a number of small, rich, disconnected bodies were found along the rift, pitching constantly northward and downward from the 1800-foot level of the Cons. California and Virginia to the 2500-foot levels of the Mexican and the Union.[8] The total production from the four mines from 1899 to 1920 was almost $7,000,000, but owing to the continuously heavy pumping expense and the uncertain management, only $568,790 was paid in dividends.[9] Meantime the assessments of the North End mines

---

[8]Much of the ore found along the northeast rift during this period averaged from $80 to $100 a ton, and was of the same character as that formerly yielded by the Con. Virginia bonanza, which led to the theory that the bonanza was mineralized by solutions uprising through the rift.

[9]The Cons. California and Virginia produced $1,500,000 (coin value) from 1899 to 1920, and paid $64,800 in dividends; the Ophir $3,900,000, and paid $262,080; the Mexican $1,145,174, and paid $161,910, and the Union $200,000, and paid $80,000.

A critical account of the methods of the Pumping Association and its broker-management was published by Whitman Symmes in 1913 in a thick pamphlet entitled *The Situation On The Comstock*. It is intended as a justification of his course as superintendent of the Association.

Symmes, before he became connected with the Comstock, wrote an article for the *Mining and Scientific Press* in which he told of "bucketing," "short selling," and the manipulation of mines and the market, as practiced in 1908. *Mining and Scientific Press*, Vol. 97, pp. 498–500, 570–576.)

amounted to nearly $5,000,000, and those of all of the mines to about $14,000,000. These ore bodies were found in blocks of ground entirely surrounded by old workings, which in the '80s had gridironed that hanging-wall country from the 1800- to the 3100-foot levels in search of another bonanza, but none of the drifts or crosscuts driven by the old-timers happened to intersect any of those small ore bodies.

During this period the Pumping Association reconditioned the Ward shaft of the Bullion and the Julia and attempted to sink below the 2525-foot point, but the heat was so great and the ground so bad that the attempt was abandoned after the expenditure of about $1,500,000. The shaft was sunk only 10 feet, and drifting north and south was stopped by intense heat and running ground. At times "the miners wore gloves to keep the ends of their finger nails from being charred, water at 170 degrees squirted from drill holes, and it became necessary to play a hose of cool water on the men as they worked."

The North End mines during this period were gradually unwatered down to 2,700 feet in depth. In 1918 the lower levels were allowed to flood again up to the 2000-foot level, and after 1920 all work was carried on above the Sutro Tunnel level.

During this period of 1900 to 1920, and for the two decades prior, the small mines and mills in Silver City produced steadily on a small scale from near-surface gold ores, maintaining a quiet well-to-do community not far above the location of the original motley camp of the early placer miners of Gold Cañon.

CHAPTER XXIX

THE BALANCE SHEET OF THE COMSTOCK

An unfavorable feature that was an integral part of the history of the Comstock was the disastrous stock gambling fever with its record of self-deprivations, broken hopes, and shattered lives. Along with this was the unsavory assessment record, often for the benefit only of the few in control of the mines. These debits have been dwelt upon in the preceding pages. The credit side of the balance sheet needs to be emphasized to the reader as he closes this history, that he may take pride in his newly won knowledge of the Comstock Lode.

The Comstock was the first silver mining camp in the United States, and its discovery brought a new era not only to California and Nevada, but to the entire West.

It lifted California out of a disheartening depression. It rejuvenated San Francisco, which in 1860 was but a ragged little town of fifty-two thousand people. In 1861 more substantial brick buildings were erected there than in all of the preceding years, nor did that growth ever cease. The opportunity for investments in the early years was limited, and nearly all of the profits from the Comstock were invested in San Francisco real estate and in the erection of fine buildings. However, the entire State shared in the benefits. California was the source of all supplies, from fruit to mining machinery, and every industry thrived. Even the money that the Californians had contributed for assessments was returned in purchases.

The discovery of the Comstock led men to look for mines throughout Nevada and in distant regions. Rich placers were found in Colorado in 1860, and soon afterward in Idaho and Montana. In Nevada, the thriving producing camps of Austin, Hamilton, Eureka, and Belmont sprang up, along with many smaller ones over the State. Mining for the first three decades in the State's history was the main industry, accompanied by the slow growth of the grazing, agricultural, and transportation industries. Mining was the economic factor that caused the separation from Utah of Nevada as a Territory, and later justified and supported statehood for Nevada.

During the Civil War the production of the Comstock mines of over fifty millions in silver and gold was a distinct aid to the

National Government. When Senator Stewart went to Washington in 1865 President Lincoln said to him: "We need as many loyal States as we can get, and the region you represent made it possible for the Government to maintain sufficient credit to continue this terrible war for the Union."

The continued production through the Bonanza days of the '70s aided in the Nation's recovery and its great industrial expansion.

EPILOGUE

The romance of the Comstock will never die. The story is an epic. It was the last stand of the California pioneers where they rose to the height of their brilliant and adventurous careers; and a robust and optimistic people throughout Nevada, many of whom were also pioneers, shared in making unforgettable history. Life was never the same for many of them in the after years, but nothing could take from them their golden memories of the Comstock Lode.

CHAPTER XXX

The South End and Middle Mines; The United Mining Co. and Comstock Merger Mines Adventures—The North End Mines, The Con. Virginia—Silver Slides into the Depression.

In the relatively quiet years following the beginning of the twentieth century, little of the excitement of earlier times remained on the Comstock Lode. Exploration activities in the deepest levels of the historic mines uniformly had resulted in failure, and the deep mines were allowed to flood after about 1885. Another attempt at exploration in the deep levels of the North End Mines was made beginning around 1900, but this venture also failed, and the lower levels again were allowed to flood in 1918. Giving up hope of finding new deep bonanza ores, Comstock operators turned their attention to recovering whatever ore remained in the upper portions of the old mines. Ore of milling grade was known to be present in stope fill and as low-grade material in the walls of earlier-mined lodes. Several of the old bonanza mines along the Lode extended their operating lives into the 1920s by mining and milling these low-grade remains.

Responding to a general increase in metal prices brought about by World War I, silver rose from $0.67 per ounce in 1916 to $1.12 per ounce in 1919. Mining technology also advanced during this time period, opening the way to larger-scale operations and their resulting lower overall costs. These factors combined to refocus attention on the Comstock and brought about the first major action on the Lode since the bonanza times of the 1870s.

In 1919, mining engineer Roy Hardy, a graduate of the Mackay School of Mines and an associate of George Wingfield, and Alex Wise, another engineer with long-standing Comstock experience, began investigation of several properties on the south end of the Lode.[1] One mine, the Imperial, was opened in 1919, and there were indications that there was plenty of ore present to justify building a mill. According to Roy Hardy, Hardy and Wise then approached the Harry Payne Whitney interests of New York with the idea of developing and opening all of the South End properties.[2] The two received authorization to go ahead with their idea and proceeded to purchase the Alpha, Imperial, Confidence, Challenge, Yellow Jacket, Crown Point, Belcher, Segregated Belcher,

---

[1] *Engineering & Mining Journal,* January 1920.

[2] Roy A. Hardy, "Reminiscence and a Short Autobiography," University of Nevada Oral History Project, 1965.

Overman, and Keystone Mines in the Gold Hill area. These properties were part of the estate of financier William Sharon and included a 5,000-foot section of the Lode along with a mill site in American Flat. In 1920 the United Comstock Mines Company was organized as a subsidiary of Harry Payne Whitney's Metals Exploration Co. by merging the companies acquired by Hardy and Wise. The principals of the company were George Wingfield, Bulkeley Wells, H. G. Humphrey, and Roy H. Elliott. Roy Hardy became resident engineer and manager in Gold Hill. Upper levels of the Gold Hill mines were reopened and sampled, and the company estimated ore reserves to be 4 million tons of $6 ore, largely in the Imperial Mine above the 400-foot level. By March 1922, a 9,585-foot tunnel had been driven from a portal in American Flat, entering the Lode at the Knickerbocker Shaft and advancing along the footwall of the Lode to the Imperial Mine at the north end of the property. A 2,500-ton-per-day cyanide plant, at the time the world's largest, and a company town, Comstock, were constructed in American Flat. A two-mile connection was made with the Virginia & Truckee Railroad, and a power line for the venture was extended from Lake Spaulding on the western slope of the Sierra Nevada.[3] Capital investment by United Comstock Mines Co. for the mine plant, mill, and associated facilities was about $5,000,000.

Production from the new facility began on September 23, 1922, and almost at once problems arose. Ore was mined using a glory-hole and caving system, not good on the Comstock with its tradition of unstable, difficult-to-control ground. The result was dilution of the ore by inclusion of large tonnages of waste, which brought down the average grade of ore going into the mill to very low grade. The ore was sticky and difficult to handle, causing costs to be higher than anticipated. Development fell behind mining needs, and the mine was unable to supply sufficient ore to run the mill at full capacity and efficiency. In addition, the price of silver dropped from $1 an ounce in 1920 to about 67 cents an ounce in 1924. It should have been no great surprise to anyone that, about this time, United Comstock's parent company, Metals Exploration Co., decided to move its interest and financial backing from Virginia City to Canada.[4] These problems proved insurmountable, and in 1924 United Comstock Mines Co. ceased its operations and sold out to the Merger Mines Co. at a considerable loss.[5]

To the north of United Comstock's property, a second large-scale mining venture got underway in 1922 when Comstock Merger Mines,

---

[3] Ibid.
[4] Ibid.
[5] Carl A. Stoddard, *Mineral Deposits of Lyon and Douglas Counties, Nevada* (Nevada Bureau Mines and Geology Bulletin 49, 1950), 29.

Inc., was formed to take over the Bullion, Best & Belcher, Savage, Gould & Curry, Exchequer, Caledonia, Chollar, Potosi, Hale & Norcross, Lady Washington, and Alta lodes.[6] These properties, referred to as the Middle Mines, covered the central section of the Comstock and extended from the United Comstock ground on the south to the Consolidated Virginia holdings on the north. Comstock Merger Mines, Inc., was a subsidiary of Gold Fields America Development Co. and was controlled by New Consolidated Gold Fields, Ltd., of London.[7] As of May 1923, the company claimed to have 4.5 million tons of reasonably assured and probable ore of $5 to $6 grade, mainly in old stope fill.[8] During much of 1923, this company engaged in reopening old stopes, sinking shaft, and conducting general underground exploration work along its section of the Lode.

In mid-1924, Comstock Merger Mines acquired all of the physical assets of the United Comstock Mines Co., and United Comstock Mines Co., with its grand plan to rejuvenate the Comstock Lode on a large scale, passed from the scene. The Merger Mines operation continued to produce from both the South End and Middle Mines until 1926. Between 1924 and 1926, the two operations removed a little less than 2 million tons of ore from the Imperial Mine; the site of the first discovery of ore on the Comstock at Gold Hill was now a deep, narrow open cut into the mountain side. Mining on the Middle Mines segment was confined to old stope fills mined underground and was done using a top-slicing method that required working down from the highest levels. Although only 45,000 tons are reported to have been mined, the resulting subsidence threatened the continued use of Virginia City's school building and the city's main street.[9] The cost and necessary mine work necessary for a successful top-slicing operation were found excessive, and on December 15, 1926, all operations were suspended indefinitely.

What went wrong? Probably a lot of things. Higher than anticipated mining and milling costs were important factors that contributed to the failure of both operations. Prior to startup in 1922, United Comstock Mines estimated that its total operating cost would be $3.00 per ton. Costs reported in 1924 were $3.35 to $3.60 per ton, and in 1926 Comstock Merger Mines reported its costs at $4.05 per ton. Falling silver prices contributed to United Comstock's problems; silver fell from $1 to 67 cents per ounce during its short operating life. The continued fall of the price of silver had less effect on the Merger Mines venture as the

---

[6] *The Mines Handbook* (1924), 1307.

[7] Ibid.

[8] Ibid.

[9] Stoddard, *Mineral Deposits of Lyon and Douglas Counties, Nevada,* 30.

price fell only another 5 cents to 62 cents per ounce by 1926. The major cause of failure of both operations, however, was overestimation of the ore value rather than increased costs.[10] The two operations, between startup in 1922 and closure in 1926, produced a combined total of 1,884,243 tons of ore yielding $7,389,487 in silver and gold, or $3.92 recovered per ton.[11] At a 90 percent mill recovery, the grade of mill-feed was about $4.35 per ton, significantly lower than the $6 per ton estimated by United Comstock or the $5.50 per ton estimated by Comstock Merger Mines. The notice of closure sent to Comstock Merger Mines workers on December 1, 1926,[12] leaves little doubt as to the causes of failure:

To All Employees:

Notice is hereby given that this company will discontinue operations sometime in December this year. The extreme irregularity of the fill ore bodies, their comparative smallness for low grade operations, great distances over which they are scattered, and the heavy character of the ground do not make for profitable operations when the average recovered grade of the old fills is only $4.30 per ton as compared with an estimated figure of $5.50. In spite of this discrepancy, it would have been possible to recover some of the investment, thanks to excellent operating results which you have obtained, if the company had not preferred to spend every cent it would make and other money raised for the purpose in searching for new ore bodies. Unfortunately, the accomplishment of the large amount of exploration work has failed to give any encouragement and the time has arrived to put the property on a salvage basis. It might have been possible to continue operations for sometime on this basis if it had not been for the recent drop in silver prices. The outlook for silver is too gloomy to justify hopes for better prices for some time to come and there is no alternative but to shut down.

T. C. Baker, General Manager

At the North End Mines during the 1920s, the Con. Virginia and its associated companies, the Mexican, Ophir, Union, and Sierra Nevada, were recovering small amounts of mill-grade ore for processing in the Mexican Mill. In 1922, for example, the Con. Virginia was mining 100 tons per day of ore averaging about $14 per ton from the 1,600-, 2,050-

---

[10] *The Mines Handbook* (1926), 1150.
[11] *The Mines Handbook* (1931), 1413.
[12] Sutro Tunnel Coalition Co. files, Gold Hill.

and 2,250-foot levels of the mine.[13] These companies too were hoping to develop large tonnages of low-grade ore in the old workings above the Sutro drainage tunnel, and the Con. Virginia was reported to contain "an abundance of low grade ore which would not pay to mine during the early days, but which now may prove profitable due to improved mining and milling practices."[14] Production from the property, however, continued only on a very small scale.

The Comstock mines produced about $600,000 in silver and gold annually for 1920 through 1922. During 1923–26, the United Comstock–Merger Mine years, annual production rose to over $2,000,000. In 1927, the year following the Merger Mines failure, production plummeted to about $271,000.

---

[13] *Engineering & Mining Journal* (1922).
[14] *The Mines Handbook* (1926), 1157.

CHAPTER XXXI

The Depression—A New Gold Price—Comstock Tunnel and
Drainage Co. and Sutro Tunnel Coalition—The Arizona
Comstock—The Crown Point—World War II and L-208.

With the onset of the Depression in 1929, silver continued its price
slide, bottoming at 28 cents an ounce in 1932. Comstock annual production fell along with silver to lows of $60,000 in 1930 and $64,000 in
1932, lower than any previous year in Comstock history since 1859.
The price of gold in the United States, maintained at $20.67 since 1837,
was increased to $35 on January 1, 1934, as part of the government's
efforts to pull the country out of the economic depression. On the Comstock, and across all of the historic mining camps of the West, the new
gold price gave ore in the ground about a 70 percent increase in value of
contained gold. Companies operating on the Comstock at this time were
embarking on major programs, as if they had somehow anticipated this
price windfall.

In 1919, the Comstock Tunnel Co., the successor to the original Sutro
Tunnel Co., reorganized as the Comstock Tunnel and Drainage Co.
with J. M. Leonard as president.[1] The new company carried with it the
provisions granted to the old Sutro Tunnel Co., whereby it received a
production royalty on ore removed from mines drained by the Sutro
Tunnel. In 1926 and 1927, the Comstock Tunnel and Drainage Co.
attempted the purchase of all of the mines formerly held by the United
Comstock–Merger Mines combination. Title to these properties was
finally acquired and the purchase effected on or about June 16, 1928.[2]
By this action, the former Middle Mines and South End Mines, a block
of ground extending from the Con. Virginia on the north to the Overman
on the south, were consolidated into one unit. During 1928, the Sutro
Tunnel Coalition, Inc., was organized as a subsidiary company to take
over and operate these holdings.

The next player to enter the Comstock scene was the Arizona
Comstock Corp. This company acquired the Savage, Chollar, Potosi,
and Hale & Norcross Mines, part of the old Middle Mines group, from
the Sutro Tunnel Coalition Co. in 1931.[3] W. J. Loring, a mine operator
with worldwide experience, was the president and general manager of
the new company. Mining was first done underground through the Hale

---

[1] *The Mines Handbook* (1922), 1169.

[2] Stockholder information release, Sutro Tunnel Coalition Co. files, Gold Hill.

[3] Sutro Tunnel Coalition Co. files, Gold Hill.

& Norcross Tunnel, but an open pit was soon started on Lode croppings on the Chollar property. Later known as the Loring Cut, this operation was at the south end of Virginia City, across the street to the west of the old Fourth Ward School. In 1933, the company began processing ore through a 150-ton flotation plant constructed near the portal of the Hale & Norcross Tunnel.[4] This was the first application of the flotation process on Comstock ores. Glowing company reports on this operation in 1936 described improvements in mill recovery (a cyanide circuit was being added to the flotation mill) and anticipated that much ore of a payable grade remained to mined from the huge open pit on the Lode outcrop.[5] By 1938, however, Arizona Comstock departed the Comstock much in the same fashion as the United Comstock and Merger Mines groups had done over ten years earlier. Citing economic difficulties brought on by the failure of the Comstock ores to respond to the flotation process, Loring resigned as president and the company went into bankruptcy. J. M. Leonard, head of the Comstock Tunnel and Drainage Co., was appointed bankruptcy trustee. During its period of operation, 1933 through early 1938, the company mined about 472,300 tons of ore containing $5.23 per ton in silver and gold. Milling operations, however, recovered only 69.9 percent of the value.[6] Assets enumerated in the bankruptcy petition included underground ore estimated at 1 to 1.5 million tons ranging from $8 to $10.50 per ton.[7] In 1939, all of the Arizona Comstock properties, which were held under option of purchase from the Comstock Tunnel and Drainage Co.'s subsidiary, the Sutro Tunnel Coalition, were sold to the Consolidated Sierra Mining & Milling Corp. of Detroit.

In the meantime, the Sutro Tunnel Coalition pursued work on its own account at the Crown Point Mine in Gold Hill. This work led to the announcement, in 1933, that 1 million tons of mill-grade ore had been blocked out in the Crown Point Mine.[8] The ore body was located directly under the historic high wooden trestle of the Virginia & Truckee Railroad. The trestle would have to be moved and the tracks relocated some 1,600 feet through a long cut-and-fill in order to mine the blocked-out ore. The railroad tracks were moved and construction completed on a 100-ton cyanide mill in July 1935.[9] Funds for mill construction and the rail detour were provided by a Reconstruction Finance Corporation loan of $160,000, the first mine loan to be granted by that fed-

---

[4] *Engineering & Mining Journal* (1933).

[5] Arizona Comstock Corp. stockholders report, October 20, 1936, NBMG files.

[6] Arizona Comstock Corp. monthly report, February 1938, Sutro Tunnel Coalition Co. files, Gold ill.

[7] *Engineering & Mining Journal* (July 1938).

[8] Ibid. (1933).

[9] Ibid. (August 1935).

eral agency. The Sutro Tunnel Coalition Co. was proud that it was able to repay the entire loan in less than three years.[10]

Comstock production pulled out of the Depression and maintained modest levels until the beginning of World War II. In October 1942, toward the end of the first year of the war, the War Production Board issued its order L-208. This edict closed all nonessential mining with the object of freeing critically needed mine workers to work in essential mines, those mining copper, lead, zinc, tungsten, and other strategic metals needed in the war effort. The Comstock silver-gold mines were considered nonessential and were ordered to close. In early 1943, silver was classified as a strategic metal. This restored the essential priority rating to developed Comstock mines and allowed work to proceed on some properties.[11] However, faced with a shortage of labor and critical supplies, mining did not flourish on the Lode during the war years. Production in 1945 dipped to a little over $26,000, a new low for the Comstock.

---

[10] Ibid. (1939).
[11] *Nevada State Journal,* March 17, 1943.

CHAPTER XXXII

A Small Post-War Revival—The Last Stamp Mill—Gerald Hartley's Tourist Mine—A Long Dry Spell.

In 1944, the Virginia City Mining Co. began mining with a power shovel on the property of the Con. Virginia. Zeb Kendal, president of the Con. Virginia, reported that the group had eight months to sample and evaluate properties that included the California and Virginia claims and the Mexican, Ophir, and Andies Mines. Roy Hardy was retained as directing engineer for the project.[1] In 1946, work was underway at the Con. Chollar Gould & Savage Co., and in 1947 it was reported that production of gold and silver was gradually increasing from the Comstock. The Con. Chollar Gould & Savage Co. was milling approximately 500 tons per day and was the leading producer on the Lode. Production, however, never returned to pre-1942 levels, and by the end of 1950 Comstock production essentially came to an end. News from the Comstock in 1952 reported that the last stamp mill remaining on the Lode would soon start processing ore from Cedar Hill. The 10-stamp mill, located at the north edge of Virginia City, began production in November at a capacity of 12 to 15 tons per day.[2] By 1953, however, the Nevada State Mine Inspector's report listed only one active mine in the Comstock district—the Occidental was being operated by Gerald Hartley's Old Comstock Mines Enterprise as an exhibition mine with no production.[3] Hartley's tourist attraction continued as the only "mine" in Virginia City through 1958. In 1959, the centennial year of the Ophir and Gold Canyon discoveries, recorded production from the Lode was only $3,720.

The 1960s marked the beginning of the longest dry spell in production on the Comstock since its discovery. With the exception of a few hundred dollars in 1960 and 1963 and about $3,000 in 1977, there was no production from the Comstock until large-scale open-pit operations began again in Gold Hill in 1980.

---

[1] *Engineering & Mining Journal* (1944).

[2] *Nevada State Journal,* July 26, 1952; *Pioche Record,* November 13, 1952.

[3] Annual Report, Nevada State Mine Inspector, 1953.

CHAPTER XXXIII

Union Pacific Railroad, Howard Hughes, Minerals Engineering, and finally, Houston Oil and Minerals; Back to Gold Hill and American Flat—The Con. Imperial Pit and Greiner's Bend—A Changing Community.

The decade of the 1960s marked a transition in Nevada's mining industry. Interest in precious metals deposits was revived in 1962 when a new type of gold deposit was discovered near Carlin in the northeastern part of the state. Later in the decade, the prices of gold and silver were released from government control and were allowed to respond to world market conditions. The excitement generated by the new gold discovery coupled with the anticipation of higher market prices led to a flurry of exploration for gold and silver throughout Nevada. This excitement spread to many of the historic mining districts in the state, including the Comstock, and in February 1968 the Natural Resources Division of Union Pacific Railroad leased a large portion of the old South End group of mines, including the Crown Point Mill, from the Sutro Tunnel Coalition.[1] In April, Howard Hughes (operating as the Hughes Tool Co.) exercised an option to buy 260 acres and four mining claims located in the American Flat area about five miles south of Virginia City.[2] The Hughes interests had been acquiring mining properties throughout central Nevada, and this purchase signaled their entry into the Comstock.

The Gold Hill property acquired by Union Pacific was essentially the same ground worked earlier by United Comstock Mines Co. in the 1920s and the Arizona Comstock Corp. in the 1930s. Union Pacific hoped to succeed where the earlier two ventures had failed by more accurate definition of ore reserves, by application of modern open-pit mining techniques, and by benefiting from anticipated higher metal prices.

Other changes that would drastically impact Nevada's mining industry also began to appear in the late 1960s. Public awareness and concern over environmental issues increased, and miners found themselves competing with other land users for the right to mine. To Nevada miners, accustomed to having mining accepted as the best possible use for the land, this was a new and difficult concept to grasp. Union Pacific encountered early warning signals almost at once when

---

[1] *Nevada State Journal,* February 14, 1968.
[2] Ibid., April 4, 1968.

exploration drilling began at Gold Hill. Local residents, many of them new to the Comstock and without ties to its mining history, began to question the effect that new, large-scale mining would have on their now-quiet community. Early in Union Pacific's examination program, it became apparent that a portion of the inferred ore zone extended under the Gold Hill–Silver City highway in the area of Greiner's Bend, and that it would be necessary to reroute the highway to gain access for mining. When drilling was completed, Union Pacific determined that at least 1.25 million tons of ore grading 0.08 ounces per ton gold and 3.5 ounces per ton silver were present.[3] This was not enough ore to proceed with mining, and Union Pacific left the Comstock without having to deal with highway relocation. The problem, however, remained for others to face later.

In 1975, Minerals Engineering Co. of Denver acquired the Gold Hill property formerly held by Union Pacific as part of a large land package comprising about 95 percent of the Comstock Lode.[4] The principals of Minerals Engineering Co., C. E. Melbye and R. J. Anctil, also had been in charge of Union Pacific Railroad's Comstock project.

In this same year, to the north along the Lode, Intermountain Exploration Co. and its Comstock partners acquired from Siskon Corp. a ten-year renewable lease on the Chollar-Potosi, Hale & Norcross, and Savage Mines. These central Comstock properties had been purchased by H. B. Chessher, Siskon's founder, in 1949. Independent engineering reports estimated that these mines contained, above the 237-foot level, 1.4 million tons of ore with an average of 0.10 ounces per ton gold and 3.6 ounces per ton silver. Below the 237-foot level, an additional 3.5 million tons of ore of this same grade were indicated. The company announced plans to mine the dumps of all three mines and to mine a 30- to 35-foot-thick ledge recently uncovered at the old Chollar (Loring) Pit across from the Fourth Ward School.[5]

In 1977, Houston Oil and Minerals Corp. acquired the New York Mine on the Silver City branch of the Lode and contracted to open access to the underground workings.[6] Houston had acquired all of the Nevada mineral holdings of Howard Hughes, including Hughes's Virginia City property. Work ceased at the New York Mine by the end of the year, but in 1978 Houston obtained the leases on all of the Gold Hill properties recently held by Union Pacific Railroad and later by Minerals Engineering Co. In July 1978, Houston announced that it had obtained the necessary permits from state and federal agencies, and ini-

[3] Ibid., January 16, 1975.
[4] Ibid.
[5] *Nevada Appeal,* October 24, 1976.
[6] Ibid., May 18, 1977.

tial operations at the Con. Imperial Mine in Gold Hill would begin within a week.[7] The company would process the ore at a mill to be constructed on American Flat near the ruins of the ill-fated United Comstock Mill of the 1920s.

Houston's plans apparently took local residents somewhat by surprise, and they were not uniformly pleased with the prospect of a large mining operation in their backyard. There was considerable local concern that Nevada's condemnation law, allowing property to be condemned for mining purposes, would be used to expedite expansion of the Gold Hill Pit. Public outlook on the new venture was summarized in two local newspaper reports. In the *Nevada Appeal* of August 2, 1978, staff writer Barbara Egbert wrote:

> Gold Hill residents feel threatened by new mining: A revival of mining on the Comstock has been first a dream, then a rumor, and now, with Houston International Minerals Corp. planning to mine, it appears to be a large-scale reality. Residents of Gold Hill, however, do not see it as a dream come true, and they may go to court to prevent the mining company from taking the community back into the last century when the mining companies ran the show and everyone else kept out of their way. Residents are not only upset about the prospect of a huge open pit in their backyards, dust and truck traffic, but are really upset about the danger to their homes and way of life. The problem is the state law of Eminent Domain that recognizes mining as of paramount interest to the state and gives mining companies the right to condemn private property it needs for mining, much in the same fashion a city or county can condemn property for public use. The Houston International Minerals Corp. lawyer stated the company had no plans to condemn property at the present time, but that they did have that right.

An item appearing in the August 9, 1978, *Reno Evening Gazette* reported:

> A Texas mining company, Houston Oil and Minerals, is giving a rebirth to the area's long-dead minerals industry, and the people living in the communities of Virginia City and Gold Hill are singularly unimpressed. People now live here for the peace and quiet and for the dollars that come from passing tourists; they regard with no small amount of concern bulldozers shoving great piles of dirt and rock around.

---

[7] *Nevada State Journal*, July 15, 1978.

The concern about condemnation of private property and destruction of historical sites was centered on the Greiner's Bend area of Gold Hill, where the Con. Imperial ore body extended under the state highway. To mine this ore would require the highway and several historic houses along it to be destroyed or relocated. In deference to Gold Hill residents living adjacent to the new pit, the company announced that its work days would consist of ten-hour shifts with no nighttime operations. By November of 1978, however, operations in the new pit were being carried out around the clock.[8] At the start of operations in mid-1978, Houston avowed no intention of using the Nevada condemnation procedures to expand its holdings at Gold Hill. In October 1979, however, Houston won a court decision allowing it to use the hundred-year-old law to take over a Gold Hill couple's private property for use in its mining operation.[9]

In February 1980, serious problems confronted the Houston operation. Massive rock slides occurred in the high wall of the open pit, and cracks started appearing in the main highway below Greiner's Bend. Excavation in the pit had to be suspended. In order to continue operations, it would now be necessary to expand the pit to the east beneath the highway. The existing road would be moved, a temporary or permanent bypass would be constructed, and the ten historic houses on the site would be purchased and removed. Since the special-use permit granted to the mining company by Storey County had restricted mining to a small area surrounding the Con. Imperial Pit, Houston now requested county permission for pit expansion and highway relocation. The two-hundred-member Save the Comstock organization opposed the mining venture, but other long-time Comstock residents pointed out that it was mining, not tourism and history, that built Nevada.[10] Following considerable debate, Storey County commissioners gave their approval to the plan. For its part, Houston International Minerals Corp. was to place $1,000,000 in a Storey County foundation plus pay the county $35,000 a year as long as it processed ore at its American Flat mill. Houston would also have to pay for relocation of State Route 342 east of Greiner's Bend.[11] None of this, however, came to pass, and in February 1981 Houston announced that it was abandoning its Greiner's Bend mining plans; the cost of relocating the state highway and the houses along it, plus the $1,000,000 the company had agreed to pay Storey County, made the enlargement of the Gold Hill pit uneconomical.[12] Crews planned to continue to mine the Con. Imperial Pit through

---

[8] *Nevada Appeal,* November 30, 1978.

[9] *Reno Evening Gazette,* October 25, 1979.

[10] *Nevada State Journal,* May 30, 1980.

[11] *Reno Evening Gazette,* October 2, 1980.

[12] *Nevada State Journal,* February 4, 1981.

the end of 1981, stopping short of Greiner's Bend, and about half of the identified ore body would be left in the ground to avoid mining in the area of controversy. Among other problems, Houston stated that it had difficulty obtaining the rights of way necessary to relocate the state highway. By mid-December, all mining activity had ended, and backfilling of the pit with waste rock was underway. The company continued to treat stockpiled ore for several months, but all operations ceased at the end of June 1982.[13] Houston had invested $25,000,000 in mine and mill construction at Gold Hill,[14] but with such a short period of operation between start of exploration in 1978 and mine closure in November 1981, not even the costs of operation were recovered.[15] Production records are unavailable for 1980 and 1981, but in 1982 recorded production was only $527,000 for the last six months of mill operation.

---

[13] *Mason Valley News,* July 2, 1982.

[14] *Las Vegas Review-Journal,* August 22, 1982.

[15] *San Francisco Chronicle,* June 26, 1982.

CHAPTER XXXIV

The Middle Mines, United Mining Corp. Goes Underground—Art Wilson Co. and the Crown Point Mill—Marshall Earth Resources on the North End.

At the time Houston was developing and mining in Gold Hill, to the north, on the Middle Mines segment of the Lode, United Mining Corp. was undertaking the first underground exploration on the Comstock in forty years.[1] With an exploration and development budget of $4,000,000, United Mining sank an exploratory shaft into the workings of the Chollar, Potosi, Hale & Norcross, Savage, and Gould & Curry Mines, and in May 1979 the company opened its New Savage Mine. United Mining estimated property reserves of 10 million tons of indicated and projected ore reserves recoverable above the 580-foot level, averaging 0.08 ounces per ton gold and 3.1 ounces per ton silver.[2] The new 320-foot ventilation and escape shaft for the New Savage Mine, opening to the bottom of the old Loring Pit, was formally dedicated on July 11, 1981, by Senator Howard Cannon.[3] The company tunneled 6,800 feet through the old Potosi and Chollar Mines to a point 435 feet under the old Fourth Ward School house on the south end of Virginia City, and by July 1, 1982, about 10,000 tons of ore had been removed from the mine.[4] Underground work stopped in January 1983, however, when the price of silver dropped. During this same month, United Mining Corp. acquired all of Houston International Minerals Corp.'s holdings on the Comstock, including the Comstock Mill on American Flat, for $10,000,000.[5] Plans were announced to open the Comstock Mill in April to process stockpiled ore from the earlier mining operation. Milling began on schedule in late April, and the New Savage Mine was reactivated in September. Underground mining continued into 1984 and, in August of that year United Mining announced plans to mine from the old Loring-Osbiston Pit and eventually to begin another open pit at the site of the Gold Hill Virginia & Truckee Railroad depot. These plans only partially came to pass. In April 1985, United Mining Corp. stated its mill was operating at a loss because of a drop in gold and silver prices, and it closed all operations, bringing to an end another

---

[1] *The Mining Record,* March 26, 1980.

[2] *Nevada Mining Association Bulletin,* April 1980.

[3] *Nevada State Journal,* July 7, 1981.

[4] *San Francisco Chronicle,* July 11, 1982.

[5] *Ely Daily Times,* January 14, 1983.

Comstock Lode mining venture. The company announced that it was searching for a buyer for the mine and mill and would sell all its assets at auction if a buyer did not come forward.[6] During its short tenure on the Comstock, United Mining drove 7,000 feet of new workings beneath Virginia City and invested $18,000,000 in the operation. The final chapter on the United Mining venture closed in 1989 when the remnants of the company property in Virginia City were auctioned.[7]

In 1988, new activity surfaced at both the north and south ends of the Lode. On the south, the Art Wilson Co. leased the Crown Point Mill from the Sutro Tunnel Coalition Co. and announced plans to mine from the Keystone Pit, located on the Silver City branch of the Lode. To the north, Marshall Earth Resources Co. began work driving a 2,000-foot tunnel from under Six Mile Canyon to intersect the Lode beneath the Con. Virginia claim.

Plans to reopen the historic Crown Point Mill rekindled the controversy over mining versus historic values that took form during the Houston Oil and Minerals venture of the early 1980s. Neighbors of the mill in Gold Hill and preservation groups acknowledged that reopening the mill would be a historic use of the site, but they also complained that dust, noise, traffic, etc., would disrupt their lives and encourage more open-pit mining. Open-pit mining, they feared, would damage the visual quality that now attracted tourists to the endangered Virginia City National Historic Landmark District. Despite objections from some residents, Art Wilson was granted permission by the Storey County commissioners to expand the Keystone Pit.[8] The Art Wilson Co. rehabilitated the Crown Point Mill, last operated in 1942 when it was shut down by War Production Board Order L-208, and planned to reopen it before the end of 1989.[9] Gold and silver suffered price drops starting about this time, however, and the Art Wilson Co. venture passed from the Lode with no production to its credit.

The Comstock was quiet in 1997. The North End Mines were still held by Marshall Earth Resources Co., but activity there was confined to drilling in the area of the old Union Mine. Official records credit no gold-silver production to the Comstock after 1985. Some production did, however, come from the Silver City branch of the Lode south of Gold Hill during this time, and other, moderately substantial production during these years came from the Flowery Lode, a parallel vein system lying over two miles east of the Comstock.

[6] *Reno Gazette-Journal,* January 18, 1986.
[7] Ibid., April 29, 1989.
[8] Ibid., April 24, 1988.
[9] *California Mining Journal,* January 1989.

CHAPTER XXXV

The Balance Sheet of the Comstock, 1997

The Comstock Lode assured its place in history by its nineteenth-century performance, and mining efforts on the Lode since 1920 can in no way detract from the earlier record. On the economic balance sheet, however, the post-1920 years were overwhelming failures. Recorded production from the Comstock (excluding production from Silver City and the Flowery Lode) for the seventy-six-year period starting in 1920 totals slightly over $37,200,000. Estimated exploration, development, and mine/mill construction expenditures for this time total at least $48,000,000 (United Comstock Mines Co., $5,000,000; Houston International Minerals, $25,000,000; United Mining Corp., $18,000,000). Expenditures by Comstock Merger Mines, Arizona Comstock Corp., Union Pacific, and other smaller companies are unknown but were presumably substantial. The Comstock, therefore, extracted at least $11,000,000 more from its mineral investors during this time than was returned to them in gold and silver.

Mining evolved during this timespan from traditional Comstock labor-intensive underground methods to large-scale surface mining using huge power shovels and haulage trucks. Virginia City residents during this same time period also evolved. Preservation of the historic remnants of the early mining camp and of the quiet lifestyle that settled upon it during the post–World War II years gained importance and came into conflict with the traditional mining interests of the district. The activity of Houston Oil and Minerals Corp. at Gold Hill between 1979 and 1982 raised some vital issues that would now forever face the mining industry. During the height of the controversy over Greiner's Bend and the historical values in Gold Hill, residents of neighboring Silver City drafted and proposed an ordinance that would require mining companies to obtain a special-use permit to mine in Lyon County. Despite strong opposition by the Nevada Mining Association and miners throughout the state, the ordinance passed. This was hailed as a victory by environmental and preservation interests, but Nevada miners viewed the ordinance as yet another burden threatening to overwhelm them and drive mining from needed mineral lands. The use of the eminent domain law in Gold Hill by Houston Oil and Minerals attracted the attention of the 1980 Nevada Legislature and led to the passage of Assembly Bill 112. The law now requires mining companies in historic districts to gain approval of the local commissioners before they can condemn property via eminent domain.

What lies ahead for the Comstock? Along the traditional Lode, the roughly three-mile-long segment of mineralized ground marked by Cedar Hill and the historic Ophir discovery site on the north and by the Gold Hill mines on the south, future surface mining may be largely out of the question. Underground mining, however, could be undertaken without impacting the historic district and could even add interest for the tourists who now fuel the area's economy.

In 1987 Tim Collins, who, as president of United Mining Corp., was the power behind the last major Comstock venture, was quoted saying, "The Comstock will always attract mining companies and hope; the next John Mackay is just over the horizon." If so, the next rise in silver and gold prices may well set the Comstock in motion again.[1]

---

[1] *Reno Gazette-Journal,* June 8, 1987.

APPENDIX

Table of Production of Comstock Mines from 1859 to January 1, 1882. Notes on Table. Comstock Production and Profits from 1859 to 1882. Table of Production of the Comstock Mines from 1882 to 1919, Inclusive.

PRODUCTION OF COMSTOCK MINES FROM 1859 TO JANUARY 1, 1882

(Silver first at coin value then market value after demonetization, February 12, 1873.)

Productive period	Tons	Yield	Per ton	Dividends	Assessments	Last dividend
Utah........		None			$1,310,000	
Sacramento & Meredith (1862–1866)	27,500	$220,000	$8.00		200,000	
Sierra Nevada (1869–1872)	63,916	625,887	8.00	$102,500		1871
Sierra Nevada (1879–1881)	9,315	326,304	35.00		4,637,000	
Union (1879–1882)	35,640	1,221,639	34.27		1,310,000	
Ophir (1859–1867)	70,771	5,294,644	74.28	1,394,400		1880
Ophir (1874–1882)	167,374	5,868,010	35.00	201,600	3,117,800	
Mexican (Old) (1859–1865)	15,000	1,500,000	100.00		100,000	
Mexican (New)		None			1,915,129	
Central (1859–1863)	6,756	500,000	74.00		159,000	
California (Old) (1860–1863)	5,800	100,000	17.07		150,000	
California (New) (1875–1882)	588,586	44,031,733	74.81	31,320,000	162,000	1879
Con. Virginia (1873–1882)	809,275	61,125,757	75.53	42,930,000	600,499	1880
Andes (1875–1878)	2,532	42,705	16.86		500,600	
Best & Belcher		None			1,144,100	
Gould & Curry (1860–1874)	314,988	15,664,162	49.76	3,826,800	3,611,000	1866*
Savage (1863–1874)	453,760	15,718,146	34.64	4,208,000	5,412,000	1869
Hale & Norcross (1865–1876)	322,549	7,927,322	24.58	1,598,000	3,810,000	1872
Chollar-Potosi (1861–1878)	605,824	16,388,367	27.10	3,579,925	3,437,502	1871
Chollar (New) (1878)		None			252,000	
Potosi (New) (1878)		None			308,000	
Bullion		None			3,872,000	
Alpha (1863–1870)	7,000	175,000	25.00		870,000	
Exchequer (1863–1870)	2,000	52,000	26.00		700,000	
Little Gold Hill Mines (Bowers, Plato, Consolidated, Piute, Rice, Eclipse, Bacon, Trench) (1859–1866)	325,000	12,000,000	36.92	3,500,000		1866

Little Gold Hill Mines (1866–1876)	160,000	3,520,000	22.00	500,000	100,000	1867
Imperial (1863–1876)	232,710	5,444,832	23.40	1,067,500	1,770,000	1869
Empire (1863–1876)	153,455	3,489,795	22.74	509,600	516,500	1867
Cons. Imperial (1876–1882)	37,787	583,012	15.43	....	1,350,000	
Confidence (1863–1870)	41,058	1,076,520	26.22	78,000	373,080	1865
Challenge (1863–1874)	7,143	200,000	28.00	....	320,000	
Yellow Jacket (1863–1874)	472,153	13,121,176	27.79	2,184,000	5,238,000	1871
Kentuck (1866–1873)	138,094	4,502,000	32.60	1,252,000	337,500	1870
Crown Point (1864–1878)	842,552	29,814,507	35.39	11,588,000	2,623,370	1875
Belcher (1863–1881)	738,171	33,813,015	45.81	15,397,200	2,419,000	1876
Seg. Belcher (1865–1872)	8,124	154,351	19.00	....	286,400	
Overman (1861–1877)	104,900	1,678,388	16.00	....	3,450,800	
Caledonia (1870–1878)	30,015	400,000	13.25	....	2,135,000	
Knickerbocker	....	None		....	564,000	
Rock Island	....	None		....	480,000	
Baltimore	....	None		....	1,015,000	
Alta (1879–1880)	463	4,808	10.38	....	1,587,600	
Justice (1873–1879)	183,174	3,369,394	18.34	....	3,361,250	
Julia	....	None		....	1,361,000	
Dayton (1862–1874)	15,000	340,000	22.66	....	859,205	
Trojan (1877–1879)	12,810	144,392	11.27	....	325,000	
Silver Hill (1863–1879)	75,000	750,000	10.00	....	1,800,900	
Woodville (1872–1875)	7,076	121,813	17.21	....	750,000	
Buckeye (1867–1874)	12,768	178,765	14.00	....	432,000	
Kossuth (1860–1870)	3,846	50,000	13.00	....	421,200	
Daney (1860–1870)	15,000	225,000	15.00	56,000	634,000	1869
Succor (1860–1875)	40,620	487,400	12.00	22,800	798,000	1872
Lady Bryan (1862–1876)	3,425	64,507	18.83	....	687,500	
Occidental (1866–1873)	20,500	410,000	20.00	20,000	355,000	1869
	7,189,430	$292,726,310	$40.72	$125,335,925	$73,929,935	

*The Gould & Curry paid a "sporadic and speculative dividend" of $48,000 in 1870.

The production and the assessments of a number of the mines are estimated in part owing to the conflicting statements.

The additional production of $27,000,000 from tailings, etc., as stated in the notes following this table, brings the total from 1859 to 1882 up to $319,726,310, or in round numbers $320,000,000, of which approximately 56 percent was silver and 44 percent gold by value.

NOTES ON COMSTOCK PRODUCTION FROM 1859 TO 1920

The foregoing statement of the production of the Comstock mines is as close an estimate as can be made. The production during the '60s and '70s is fairly dependable, as nearly all of the annual reports of the companies have been found, but, even there, some allowances have to be made. The statement during the '70s is practically correct, including gold and silver values, with silver estimated at its coin value. Prior to 1866 none of the mining companies reported the gold and silver values separately. Few of the mines made regular reports after 1886.

When Lord endeavored to state the year-by-year production from 1859 to 1880 he was confronted with estimates from different sources, which varied in every year. (See Lord on pp. 416, 417.) His estimate of $306,000,000 is about $14,000,000 too low.

After 1880 the statements and reports of the U. S. Geological Survey, the Director of the Mint, and U. S. Mineral Resources are only partly dependable, as shown by the fact that they often contradict themselves. There was much careless work done for the Director by local investigators. The County Assessor's figures are the most reliable, but they differ among themselves at times.

The inclusion of a large tonnage of tailings and low-grade ore after 1900 materially reduced the value per ton of the $7,000,000 worth of ore produced during the succeeding fifteen years by the North End mines when they were pumped out to the 2500-foot level.

The Government now presents us with an elaborate and impressive tabulation of production from 1859 to 1938, giving the gold and silver values for each year. But the relative production of gold and silver cannot be accurately stated except during the '70s, and in small part later, and the statement was but an estimate of the Government offices. Its value may be judged by the fact that the elaborate report of the Director of the U. S. Mint for 1887, at page 230, stated the year-by-year values in gold and silver from 1859 to 1887 in which the gold appeared largely in excess. In the 1889 report, at pages 169 and 170, the table was revised, and silver, which is usually estimated at $1.2929 per ounce, appears as greatly in excess over gold. The report also arbitrarily adds about 900,000 tons to production from 1860 to 1867, thereby reducing the average value per ton during those years to about $25, when, in fact, it was about $60.

An exact statement of the production of the Comstock mines, including returns from tailings, cannot be made. The differences amount to several millions of dollars. At the very most the production did not exceed $400,000,000, and $396,000,000 would be nearer the mark. It is quite bold to state the production in any year in exact hundreds of tons or hundreds of dollars, unless it is understood to be only as close an approximation as can be made, the reliability depending upon the depth of research, the ability and the integrity of the compiler.

The production up to 1871 was approximately 60 percent silver and 40 percent gold. Thereafter the Crown Point and Con. Virginia bonanzas changed that by returning a slight excess in favor of gold. After 1880 silver fell rapidly in value, and the total value production of the Comstock may be stated at 55 percent silver and 45 percent gold.

COMSTOCK PRODUCTION AND PROFITS FROM 1859 TO 1882

One would expect to find reasonably accurate statements of the assessments and the production of Comstock mines in official Government reports, but the figures given by Lord and by Becker up to 1881 are so imperfect that a laborious search has been made to get at the facts. It would have been a much simpler task when they were writing, as all of the information was then available.

Becker's table of the production of individual mines on page 10, which he says was prepared by Lord "with great pains," is imperfect and misleading, as it fails to mention some of the producers and gives the yield of others only during certain periods. However, he brings the total to something near the fact by adding $23,598,947.02 for "estimated additional production, chiefly in earlier years," in order to make it agree with Lord's table on page 416. Both of their estimates are based on silver at $1.2929 per ounce after 1872, instead of sale value, which increases the yield of the bonanza mines and of the Crown Point and of the Belcher by about $10,000,000.

Becker's estimate of $3,765,000 from tailings "worked by various private mills" is fully $20,000,000 too low. The Bonanza Firm alone produced about $12,000,000 from tailings up to 1881, and Sharon and Jones as much more.

Lord's list of assessments (pp. 419–423) was copied from "The Daily Stock Report and Assessment Guide," of June 1, 1880, which published such lists for years. Only the stocks listed on the Exchange at the time were included. The companies that had been stricken from the Board, and those that had passed out of existence, were omitted. The constant effort of the Exchange was to minimize assessments. Lord's list, therefore, includes only 103 companies, whereas, by actual count, there were over 400 others that operated in the district between 1859 to 1882, whose omitted assessments aggregated many millions.

After the fashion of the Stock Exchange journals, Lord draws the following comparison (p. 353) between dividends and assessments to June, 1880: Assessments, $62,000,000; dividends and profits, $118,000,000; net of operations, $56,000,000; but he omits $5,157,025 in dividends paid by the Chollar-Potosi, the Imperial, and the Empire, together with many millions in private profits, and fails to include the additional assessments levied by many other stock companies, estimated at $14,000,000. Furthermore, the assessments of a number of the mines he mentions are underestimated, while $1,735,800 levied by the Imperial and the Empire, and $2,753,502 levied by the Chollar-Potosi are omitted altogether. He omits also the assessments of a number of the larger mines then operating, such as the Rock Island, the Knickerbocker, and the Florida. Lord's list, and his comparison between assessments and dividends, therefore, has little value.

Lord's estimate of private profits (p. 353) is $2,000,000, whereas the Bonanza Firm alone made $9,070,728 from milling contracts, including profits from tailings, and Sharon and associates, and Jones and associates, in like manner, gathered in an additional $10,000,000. (It was not considered "good form" at that time to refer to private profits.) The total of private profits was not less than $20,000,000. Expenditures by nonboard companies (referred to as "private companies"), and by individuals, from which no returns were had, may be placed at $11,000,000.

The preceding production table from 1859 to 1882 is based upon official mine reports, in the main, and makes liberal allowances for minor properties whose production, etc., are not accurately known.

Production of The Comstock Mines From 1882 to 1919, Inclusive*

Year	Tons	Yield
1882	90,173	$1,675,002
1883	125,914	1,995,035
1884	188,368	2,620,751
1885	238,942	2,952,449
1886	268,979	3,429,663
1887	223,682	3,820,120
1888	271,598	5,665,335
1889	284,560	5,525,782
1890	286,875	4,062,019
1891	186,394	2,733,219
1892	133,671	1,963,333
1893	109,780	1,872,104
1894	97,049	1,281,468
1895	63,558	914,789
1896	39,240	804,088
1897	17,850	610,014
1898	10,766	410,015
1899	6,780	389,161
1900	12,109	299,179
1901	35,948	719,679
1902	43,027	392,676
1903	31,900	265,941
1904	45,360	812,179
1905	39,993	553,433
1906	36,281	591,980
1907	16,551	299,956
1908	64,943	814,752
1909	87,061	727,494
1910	109,152	553,033
1911	117,202	1,330,242
1912	118,283	1,083,288
1913	193,506	1,181,568
1914	100,016	350,559
1915	82,236	285,809
1916	49,908	448,002
1917	48,597	436,728
1918	52,796	659,321
1919	43,081	635,860
	3,972,129	$55,166,026

*Compiled for the author by Mr. B. F. Couch, Secretary of the Nevada State Bureau of Mines, from statements rendered by producers under the net-proceeds tax law of the State.

Appendix II

The Comstock Production Record, 1920–1996

Year	Lode Material, in tons	Gold Placer, oz.	Gold Lode, oz.	Gold Value, in dollars	Silver Placer, oz.	Silver Lode, oz.	Silver Value, in dollars	Copper Pounds	Copper Value, in dollars	Lead Pounds	Lead Value, in dollars	Total Value, in dollars
1920	58,494		15,576	321,982		324,744	353,971					675,953
1921	64,601		15,714	324,827		263,931	263,931					588,758
1922	138,752		14,436	298,409		329,644	329,644					628,053
1923	584,237		46,286	956,818		1,357,559	1,113,198	912	134			2,070,150
1924	547,428		53,635	1,108,743		1,616,692	1,083,184	720	94			2,192,021
1925	474,325		57,902	1,196,946		1,350,156	937,008	838	119			2,134,073
1926	481,767		57,144	1,181,277		1,534,116	957,289	3,390	475			2,139,041
1927	81,827		11,402	235,690		62,051	35,183	1,100	144			271,017
1928	76,441		8,729	180,442		37,769	22,042	5,709	822	172	10	203,316
1929	11,244		5,315	109,879		32,399	17,269	2,611	460			127,608
1930	6,198		2,436	50,352		25,296	9,739					60,091
1931	17,931		4,274	88,358		26,074	7,561	1,390	126	6,619	245	96,290
1932	5,788		2,886	59,657		13,908	3,922	215	14	1,799	52	63,645
1933	32,193	2	5,056	104,567		84,026	29,409	580	39			134,015
1934	155,563	354	16,312	582,460	99	296,405	191,679	1,981	158	269	10	774,307
1935	225,929	10	17,910	627,211	7	333,276	239,547	1,627	135	1,347	54	866,947
1936	349,605	56	27,922	979,230	127	474,755	367,796	10,000	920	2,000	92	1,348,038
1937	486,328	29	35,913	1,257,970	38	522,174	403,931	6,000	726			1,662,627
1938	397,415	99	39,182	1,374,835	131	583,514	377,306	4,000	392	4,000	184	1,752,717
1939	310,763	12	33,629	1,177,435	15	415,569	282,093	2,000	208	200	94	1,459,830

Year										Total
	ԾԾ,Ӏ31	16	45,722	1,600,830	50	492,123	349,990			1,950,820
1941	331,502	6	42,975	1,504,335	48	539,010	383,330			1,887,665
1942	281,616	12	33,150	1,160,670	38	298,654	212,403			1,373,073
1943	96,895	28	5,225	183,855	80	123,293	87,732			271,587
1944	56,157		3,655	127,925		44,065	31,335			159,260
1945	1,284	5	656	23,135	42	4,604	3,304			26,439
1946	117,582	12	5,419	189,665		50,854	41,090	6,000	654	231,409
1947	127,036		5,016	175,980	175	68,446	62,102			238,082
1948	166,574		11,591	405,685		176,882	160,087	100	18	565,790
1949	243,233		18,540	648,900		233,705	214,515	400	63	863,478
1950	125,858		9,691	339,185		108,944	98,600	100	14	437,799
1951	2,038		267	9,345		3,512	3,179	100	17	12,541
1952	14		10	350		8	7			357
1953	204		143	5,005		143	129			5,134
1954	3,752		405	14,175		7,169	6,488			20,663
1955	6,086		314	10,990		10,676	9,680			20,670
1956	7,435		873	30,555		6,602	5,975			36,530
1957	4,240		548	19,180		4,493	4,067			23,247
1958	4,501		549	19,215		8,506	7,698			26,913
1959	669		78	2,730		1,094	990			3,720
1960	30		6	210		8	7			217
1961	No recorded production									-
1962	200		12	420		17	18			438
1963	85		21	735		112	143			878
1964-1976	No recorded production									-
1977	20		6	890		450	2,079			2,969
1978	No recorded production									-
1979	No recorded production									-
1980	No recorded production									-

Appendix II

The Comstock Production Record, 1920–1996

(continued)

Year	Lode Material, in tons	Gold Placer, oz.	Gold Lode, oz.	Gold Value, in dollars	Silver Placer, oz.	Silver Lode, oz.	Silver Value, in dollars	Copper Pounds	Copper Value, in dollars	Lead Pounds	Lead Value, in dollars	Total Value, in dollars
1981					No recorded production							-
1982	108,851		818	307,494		27,606	219,468					526,962
1983	40,000		3,600	1,526,400		124,000	1,418,560					2,944,960
1984	320,000		12,000	4,327,896		250,000	2,035,165					6,363,061
1985-1996					No recorded production							-
TOTAL	6,911,822	641	672,950	24,852,843	850	12,269,034	12,383,843	43,073	4,966	23,106	1,507	37,243,159

Index

Supplementary Index

for new material, pp. 291-308